HOLT
ENVIRONMENTAL SCIENCE

REVIEW AND CRITICAL THINKING WORKSHEETS WITH ANSWER KEY

This book was printed with soy-based ink on acid-free recycled content paper, containing 10% POSTCONSUMER WASTE.

HOLT, RINEHART AND WINSTON

Harcourt Brace & Company

Austin • New York • Orlando • Atlanta • San Francisco • Boston • Dallas • Toronto • London

TO THE TEACHER

This *Review and Critical Thinking Worksheets with Answer Key* booklet is designed to provide you with a tool for reviewing the material covered in the textbook, and to give students an opportunity to apply their knowledge to critical thinking problems. The booklet contains blackline master Chapter Review and Critical Thinking worksheets for each chapter of the Pupil's Edition of *Holt Environmental Science*. An Answer Key is provided in the back of the booklet.

Each set of chapter worksheets contains six to nine pages. The first two to four pages are devoted to a review of the chapter material and contain matching questions, concept mapping, multiple choice questions, and short answer questions. The Critical Thinking worksheets for each chapter include some or all of the following sections: Interpreting Observations, Agree or Disagree, Refining Concepts, and Reading Comprehension and Analysis. You may choose to use the Critical Thinking worksheets either together or individually, and you may also wish to use them as assessment tools, as part of a portfolio, or as an addendum to the regular Chapter Test.

Photography Credits
Front Cover: water testing, Tom Stewart/The Stock Market; erosion, Carr Clifton/Minden Pictures; Ganges Delta, World Perspectives/Tony Stone Images; ferns, Pat O'Hara/Tony Stone Images; cheetah, Jeff Hunter/Image Bank; clouds, Robert Stahl/Tony Stone Images; windmills, Lester Lefkowitz/The Stock Market; lungs, Photo Researchers, Inc.; contour plowing, D. Wigget/Natural Selection; sunflower, Jim Brandenburg/Minden Pictures; highways, Jose Fuste Raga/The Stock Market

Back Cover:
Pat O'Hara/Tony Stone Images

Title Page:
Jim Brandenburg/Minden Pictures

Art Credits
All art, unless otherwise noted, by Holt, Rinehart & Winston
Page 9, Thomas Kennedy.

Printed in the United States of America
ISBN 0-03-053844-0

3 4 5 6   862   03 02 01 00

CONTENTS

CHAPTER REVIEW

CHAPTER
1

ENVIRONMENTAL SCIENCE: A GLOBAL PERSPECTIVE

Matching   Match each example in the left column with the appropriate term from the right column.

_____ **1.** see a lizard basking in the sun

_____ **2.** measure body temperatures of lizards before and after basking

_____ **3.** graph and analyze results

_____ **4.** give a talk on why lizards bask

_____ **5.** state that lizards bask in the sun to warm themselves

a. hypothesis

b. observation

c. communicating results

d. organizing and interpreting data

e. experiment

Concept Mapping

6. Based on the graph shown below, are the depicted environmental resources—coal, sunlight, and trees—renewable or nonrenewable at current rates of use?

_____

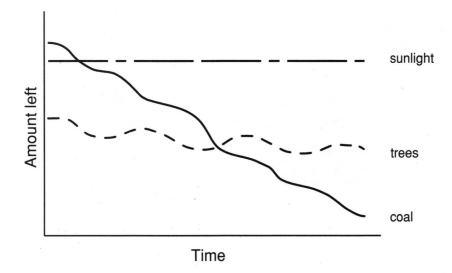

Graphing Connection   Read the scenarios below and decide which kind of graph—line graph, pie chart, or bar graph—would best display the data. Write the name of the graph and the answers to any questions on the lines below the data tables.

1. Ted suspects that his shower has been losing pressure over the last month and wonders if his pipes are getting clogged with debris. He decides that every morning for the next week he will measure how long it takes to fill a 5 L bucket with the water going full blast. Decide which type of graph is most appropriate and whether Ted's observations support his hypothesis.

Day	Time to fill 5 L bucket (sec)
1	25
2	26
3	28
4	29
5	30
6	30
7	31

_____

_____

2. Jill and Jerry have a vegetable farm. They are trying to decide which of two fields is better for growing zucchini and other types of squash. One field is in the shade for about half the day and the other is in full sun all day. Below are their July yields from each field. Decide which type of graph is most appropriate, and which field provides better yields.

Vegetable	Shady field	Sunny field
Zucchini	240 kg	400 kg
Acorn squash	70 kg	100 kg
Butternut squash	190 kg	290 kg

_____

_____

CHAPTER 1 REVIEW, CONTINUED

Short Answer   Write the answers to the following questions in the spaces provided.

1. Explain the two different meanings of science, and provide an example of each meaning.

    _____

    _____

    _____

    _____

    _____

    _____

    _____

2. Name and describe three human activities that affect the environment.

    _____

    _____

    _____

    _____

    _____

    _____

3. Explain how gathering information contributes to making sound environmental decisions.

    _____

    _____

    _____

    _____

    _____

    _____

CRITICAL THINKING WORKSHEET

CHAPTER 1

ANALOGIES

Mark the letter of the pair of terms that best completes the analogy shown. An analogy is a relationship between two pairs of words or phrases written as a:b::c:d. The symbol : is read *is to,* and the symbol :: is read *as*.

Example
keyboard : type ::

_____ **a.** plane : land

_____ **b.** dog : eat

√ **c.** scissors : cut

_____ **d.** rock : hard

_____ **e.** grass : green

1. to observe : to experiment ::

_____ **a.** to repeat : to examine

_____ **b.** to test : to study

_____ **c.** to audition : to perform

_____ **d.** to watch : to manipulate

_____ **e.** real world : computer

2. pollution : poison ::

_____ **a.** industrial : natural

_____ **b.** gas : liquid

_____ **c.** cyanide : smog

_____ **d.** byproduct : intentional product

_____ **e.** harmful to environment : toxic to
         organisms

3. values : environmental decision ::

_____ **a.** math : economics

_____ **b.** values : scientific

_____ **c.** consequences : information

_____ **d.** weather : umbrella

_____ **e.** budget : purchase

4. biosphere : atmosphere ::

_____ **a.** organisms : weather

_____ **b.** thick : thin

_____ **c.** biology : physics

_____ **d.** concentrated : diffuse

_____ **e.** maybe : definitely

5. to sustain : to consume ::

_____ **a.** to breathe : to eat

_____ **b.** to support : to deplete

_____ **c.** to manage : to kill

_____ **d.** to generalize : to specialize

_____ **e.** to grow : to die

6. observation : hypothesis ::

_____ **a.** room is hot : like heat

_____ **b.** feel hot : like heat

_____ **c.** like heat : room is hot

_____ **d.** feel hot : have a fever

_____ **e.** have a fever : feel hot

7. ecology : environment ::

_____ **a.** organism : ecosystem

_____ **b.** interactions : surroundings

_____ **c.** animals : plants

_____ **d.** theory : science

_____ **e.** biosphere : atmosphere

CRITICAL THINKING WORKSHEET

CHAPTER
1

INTERPRETING OBSERVATIONS

Read the following scenario, and answer the questions that follow.

All summer long the weather has been very hot and dry. To conserve water, your family has not been watering the plants in your front yard. The grass is getting more yellow every day, and the bushes and flowers look pretty pathetic. However, you notice that one plant is thriving. This plant is growing in the shade of a large bush near the front door.

1. Why do you think this one plant is thriving? State your answer as a hypothesis.

_____

_____

_____

_____

2. How could you test your hypothesis?

_____

_____

_____

_____

_____

3. What other kinds of information might you gather to help you explain this plant's success?

_____

_____

_____

_____

_____

_____

_____

_____

CRITICAL THINKING WORKSHEET

AGREE OR DISAGREE

Agree or disagree with the following statements, and support your answer.

1. Science is either pure or applied.

   _____

   _____

   _____

   _____

   _____

2. Students who want to be scientists should only study science; literature and the arts have little bearing on their work.

   _____

   _____

   _____

   _____

   _____

3. Most people from developing countries have values and priorities very different from those of most people from developed countries.

   _____

   _____

   _____

   _____

   _____

CHAPTER 1

READING COMPREHENSION AND ANALYSIS I

Read the following passage, and answer the questions that follow.

When you can measure what you are speaking about, and express it in numbers, you know something about it; but when you cannot measure it, when you cannot express it in numbers, your knowledge is of a meager and unsatisfactory kind: it may be the beginning of knowledge, but you have scarcely, in your thoughts, advanced to the stage of *science*.

Lord Kelvin*

1. What step in the scientific method does the author's point refer to? Explain your answer.

_____

_____

_____

_____

2. What assumptions might the author have about the nature of knowledge and science?

_____

_____

_____

_____

3. Do you agree with the author's point? Relate your answer specifically to the definition and scope of environmental science.

_____

_____

_____

_____

_____

*From a speech by William Thomson (Lord Kelvin) in 1891.

Name _____ Class _____ Date _____

READING COMPREHENSION AND ANALYSIS II

Read the following passage, and answer the questions that follow.

In the end, we will conserve only what we love, we will love only what we understand, we will understand only what we are taught.

Baba Dioum*

1. Which do you care about more—a park near your home or an area of desert in Australia?

_____

_____

2. Does your degree of familiarity with these places influence your answer? Explain your answer.

_____

_____

_____

3. Think of something you care about very much that your classmates may be unfamiliar with, such as a person, a pet, or a special place you like to go. Do you think your classmates would care more if they knew what you know?

_____

_____

_____

4. How do the author's ideas relate to the study of environmental science?

_____

_____

_____

_____

*From a speech by Baba Dioum in 1968 for the International Union for Conservation of Nature and Natural Resources, New Delhi, India. Copyright © 1968 by **Baba Dioum.** Reprinted by permission of the author.

Name _____ Class _____ Date _____

CHAPTER
2

LIVING THINGS IN ECOSYSTEMS

Matching   Match each example in the left column with the appropriate term from the right column.

_____ **1.** a single tiger shark

_____ **2.** the plankton, fish, shellfish, and other organisms, plus the water, temperature, currents, and nutrients in Lake Superior

_____ **3.** sunflowers in a field

_____ **4.** all people

_____ **5.** all of the termites, ants, beetles, mosses, and other organisms inside a rotting tree stump

a. species

b. ecosystem

c. population

d. community

e. organism

Concept Mapping

6. Demonstrate how the organisms shown below might interact with one another. First, draw a line from each organism to the other organism with which it might interact. Then write the name of the appropriate interaction—*predation, competition, commensalism, parasitism,* or *mutualism*—on the line connecting the organisms.

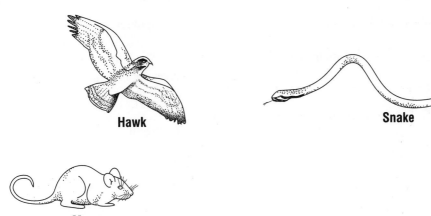

CHAPTER 2 REVIEW, CONTINUED

Multiple Choice   In the space provided, write the letter of the word or statement that best answers the question or completes the sentence.

_____ 1. Which is NOT a biotic factor in an ecosystem?

    a. weeds

    b. sand grains

    c. snakeskins

    d. mushrooms

_____ 2. Which of the following is an example of a parasite?

    a. a toadstool on a tree trunk

    b. a cancer cell in your liver

    c. a bee stinger in your arm

    d. a flu virus in your lung

_____ 3. Natural selection is not responsible for the evolution of

    a. the taste of mint leaves.

    b. the shape of a river.

    c. your ability to sweat.

    d. the shape of a fly's eye.

_____ 4. Predators _____ kill their prey.

    a. always

    b. sometimes

    c. never

    d. like to

_____ 5. Which of the following is an example of a species?

    a. German shepherds

    b. birds

    c. polar bears

    d. soil arthropods

_____ 6. Examples of communities include all of the following EXCEPT the

    a. organisms in a pond.

    b. organisms in a fish tank.

    c. microbes living on you.

    d. dogs on your block.

_____ 7. A panda bear might be all of the following EXCEPT a

    a. parasite.

    b. competitor.

    c. mutualist.

    d. predator.

_____ 8. The "co-" in coevolution, as in cooperation, means

    a. apart.

    b. together.

    c. two.

    d. predator-prey.

CHAPTER 2 REVIEW, CONTINUED

Short Answer   Write the answers to the following questions in the spaces provided.

1. Explain the relationship between natural selection and evolution.

   _____

   _____

   _____

   _____

   _____

   _____

   _____

2. Explain the difference between a community and a population. Provide a specific example of each.

   _____

   _____

   _____

   _____

   _____

   _____

   _____

   _____

3. Define *competition,* and give three examples not listed in the book.

   _____

   _____

   _____

   _____

   _____

   _____

CRITICAL THINKING WORKSHEET

ANALOGIES

Mark the letter of the pair of terms that best completes the analogy shown. An analogy is a relationship between two pairs of words or phrases written as a:b::c:d. The symbol : is read *is to,* and the symbol :: is read *as.*

Example

keyboard : type ::

_____ **a.** plane : land

_____ **b.** dog : eat

√ **c.** scissors : cut

_____ **d.** rock : hard

_____ **e.** grass : green

1. Biotic factors : organisms ::

_____ **a.** abiotic factors : plants

_____ **b.** abiotic factors : soil

_____ **c.** biotic factors : physical environment

_____ **d.** ecosystem : community

_____ **e.** plants : water

2. Organism : habitat ::

_____ **a.** dog : niche

_____ **b.** giraffe : beach

_____ **c.** bird : ecosystem

_____ **d.** turtle : pond

_____ **e.** beetle : population

3. Mutualism : cooperative ::

_____ **a.** evolution : coevolution

_____ **b.** predator : prey

_____ **c.** commensalism : harmful

_____ **d.** parenthood : nurturing

_____ **e.** parasite : host

4. Population : community ::

_____ **a.** individual : population

_____ **b.** species : organism

_____ **c.** habitat : niche

_____ **d.** biotic : abiotic

_____ **e.** organism : community

5. Natural selection : evolution ::

_____ **a.** key : keyhole

_____ **b.** rain : erosion

_____ **c.** ground : landing

_____ **d.** one : many

_____ **e.** struggle : cooperate

6. Low : high ::

_____ **a.** evolution : coevolution

_____ **b.** fall : winter

_____ **c.** sun : earth

_____ **d.** niche : habitat

_____ **e.** extinct : plentiful

7. Evolution : coevolution ::

_____ **a.** chicken : egg

_____ **b.** dance alone : dance with partner

_____ **c.** ice skate : play hockey

_____ **d.** mirror : image

_____ **e.** playing checkers : playing chess

Name _____ Class _____ Date _____

INTERPRETING OBSERVATIONS

Read the following scenario, and answer the questions that follow.

The Venus' flytrap is a species of plant that captures flies by trapping them between hinged leaves that snap shut when a fly lands on them. Flies are attracted to the leaves because they smell like food, but it is the flies who end up being eaten!

Imagine that 1 million years ago seeds of a Venus' flytrap were blown to an island. The island had three species of flies that each lived in different areas of the island and each ate different foods. In the year 2000, scientists exploring the island discover two new species of fly-eating plants as well as the Venus' flytrap.

1. What has happened?

_____

_____

_____

_____

2. Scientists call this type of evolution *adaptive radiation*. Based on the words in this term and on the scenario above, what do you think this term means?

_____

_____

_____

_____

_____

3. Suppose that the scientists only found one new species of flytrap. What might you conclude?

_____

_____

_____

_____

_____

_____

CRITICAL THINKING WORKSHEET

AGREE OR DISAGREE

Agree or disagree with the following statements, and support your answer.

1. An ecosystem can be viewed as a host that is parasitized by the organisms that live in it.

_____

_____

_____

_____

2. The ability to unconsciously regulate your body temperature is an adaptation.

_____

_____

_____

_____

_____

3. An organism can engage in only one type of interaction (i.e. predation, mutualism, etc.) with another organism at one time. Provide examples to support your viewpoint.

_____

_____

_____

_____

_____

_____

_____

_____

14

REFINING CONCEPTS

The statements below challenge you to refine your understanding of concepts covered in the chapter. Think carefully, and answer the questions that follow.

1. Although there are many predators on the African savanna, none plays exactly the same role as the lion. Can any two species occupy exactly the same niche? Why or why not?

_____

_____

_____

_____

2. Biologist Bob thinks that over time a parasite can influence the evolution of its host species. Do you think he is right? Justify your answer.

_____

_____

_____

_____

_____

_____

3. One characteristic of a population is that organisms must have a reasonable chance of mating with each other. Are two wild roses separated by a wide road part of the same population? Defend your answer.

_____

_____

_____

_____

_____

CRITICAL THINKING WORKSHEET

READING COMPREHENSION AND ANALYSIS I

Read the following passage and answer the questions that follow.

Diana lives next to a large area of undeveloped oak forest. While playing in the woods with friends, she notices three species of lizards: a large, ground-dwelling species, a medium-sized species that lives on tree trunks, and a small species that lives on tree branches. All three species eat whatever small animals they can capture, usually insects.

The oak trees have recently been damaged by a type of wasp whose larvae feed on oak leaves. To save the trees, the county decides to spray the forest from the air with an insecticide. The insecticide usually does not harm lizards, but lizards can die if they eat too many poisoned wasp larvae.

1. Which of the lizards are least likely to be harmed by the insecticide? Why?

   _____

   _____

   _____

2. Are the lizards competing for the wasp larvae? Explain your answer.

   _____

   _____

   _____

   _____

3. The county sprays the trees on a still summer day. A week after the spraying, Diana learns that ducks living in a pond just up the mountain have become sick. Diana thinks she knows why. What might Diana be thinking? Is her reasoning reasonable?

   _____

   _____

   _____

   _____

   _____

   _____

Name _____ Class _____ Date _____

CHAPTER
2

READING COMPREHENSION AND ANALYSIS II

Read the following passage, and answer the questions that follow.

The wolf is tied by subtle threads to the woods he moves through. His fur carries seeds that will fall off, effectively dispersed, along the trail some miles from where they first caught his fur. And miles distant is a raven perched on the ribs of a caribou the wolf helped kill ten days ago, pecking like a chicken at the decaying scrap of meat. A smart snowshoe hare that eluded the wolf and left him exhausted when he was a pup has been dead for a year now, food for an owl. The den in which he was born one April evening was home to porcupines last winter.

Barry Holstun Lopez*

1. What ideas from Chapter 2 does the author deal with in this passage?

_____

_____

_____

2. Explain what the author means by the wolf being "tied by subtle threads to the woods he moves through."

_____

_____

_____

3. Scientists have long battled over the idea of niches. Do niches exist? What are they? How should we think of them? Given what you have learned, explain how the niche concept is useful and how it is not useful in helping you understand an organism's way of life.

_____

_____

_____

_____

_____

*From *Of Wolves and Men* by Barry Holstun Lopez. Copyright © 1978 by Barry Holstun Lopez. Reprinted by permission of ***Sterling Lord Literistic, Inc.***

CHAPTER REVIEW

HOW ECOSYSTEMS WORK

Matching   Match each example in the left column with the appropriate term from the right column.

_____ **1.** herbivore

_____ **2.** carnivore

_____ **3.** producer

_____ **4.** omnivore

_____ **5.** decomposer

a. oak tree

b. raccoon

c. spider

d. mushroom

e. aphid

Concept Mapping

6. Draw an energy pyramid using the following organisms found in the temperate forest of North America: shrubs, trees, bears, insects, woodpecker, hawk, rabbit, cougar, and deer. Beside your diagram, explain why you placed these animals as you did.

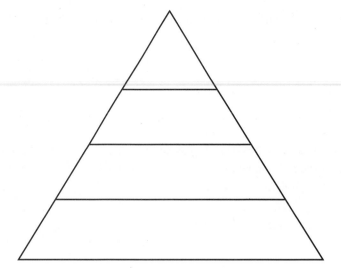

CHAPTER 3 REVIEW, CONTINUED

Short Answer Write the answers to the following questions in the spaces provided.

1. Explain how cellular respiration fits into the carbon cycle.

_____

_____

_____

_____

_____

_____

_____

_____

2. Describe the role of decomposers in the nitrogen cycle.

_____

_____

_____

_____

_____

_____

3. Describe the sequence of events that takes place after an agricultural field is abandoned. Do both primary and secondary succession occur? Why or why not?

_____

_____

_____

_____

_____

_____

_____

CRITICAL THINKING WORKSHEET

CHAPTER 3

ANALOGIES

Mark the letter of the pair of terms that best completes the analogy shown. An analogy is a relationship between two pairs of words or phrases written as a:b::c:d. The symbol : is read *is to*, and the symbol :: is read *as*.

Example
keyboard : type ::

_____ **a.** plane : land

_____ **b.** dog : eat

___√___ **c.** scissors : cut

_____ **d.** rock : hard

_____ **e.** grass : green

1. producer : consumer ::

_____ **a.** car : driver

_____ **b.** factory : shopper

_____ **c.** deer : wolf

_____ **d.** photosynthesis : decomposition

_____ **e.** plants : water

2. herbivores : omnivores ::

_____ **a.** photosynthesis : respiration

_____ **b.** elephant : ocean

_____ **c.** fruit : bird

_____ **d.** deer : bear

_____ **e.** bacteria : population

3. plant : terrestrial ecosystem ::

_____ **a.** evolution : time

_____ **b.** predator : prey

_____ **c.** algae : respiration

_____ **d.** sea urchin : marine ecosystem

_____ **e.** parasite : host

4. lake : water cycle ::

_____ **a.** coal deposit : carbon cycle

_____ **b.** species : organism

_____ **c.** rain : clouds

_____ **d.** biotic : ecosystem

_____ **e.** organism : community

5. oxygen : cellular respiration ::

_____ **a.** cup : saucer

_____ **b.** carbon dioxide : photosynthesis

_____ **c.** plants : adaptation

_____ **d.** needle : thread

_____ **e.** disk : computer

6. climax forest : clear-cut forest ::

_____ **a.** plants : animals

_____ **b.** food web : food chain

_____ **c.** sun : fire

_____ **d.** ecosystem : habitat

_____ **e.** full : empty

7. trophic level : energy ::

_____ **a.** car : fuel

_____ **b.** food chain : food web

_____ **c.** grade : demerits

_____ **d.** energy pyramid : organisms

_____ **e.** taxes : income

Name _____ Class _____ Date _____

REFINING CONCEPTS

The statements below challenge you to refine your understanding of concepts covered in the chapter. Think carefully, and answer the questions that follow.

1. Do you think the distinction between the concepts of primary and secondary succession is useful? Explain your answer.

_____

_____

_____

_____

_____

_____

_____

_____

2. If the sun were to burn out, explain what you think would happen to the water, carbon, and nitrogen cycles.

_____

_____

_____

_____

_____

_____

_____

_____

_____

_____

Name ______________________ Class ______________ Date ________

CRITICAL THINKING WORKSHEET

CHAPTER 3

Agree or Disagree

Agree or disagree with the following statements, and support your answer.

1. There would be no life on Earth without the sun.

2. Our activities do not upset the carbon cycle.

3. Urban development negatively affects the water cycle.

Name _____ Class _____ Date _____

READING COMPREHENSION AND ANALYSIS I

Read the following passage, and answer the questions that follow.

This thumbnail sketch of land as an energy circuit conveys three basic ideas:
1) that land is not merely soil;
2) that the native plants and animals kept the energy circuit open; others may or may not;
3) that man-made changes are of a different order than evolutionary changes, and have effects more comprehensive than is intended or foreseen.

These ideas, collectively, raise two basic issues:

Can the land adjust itself to the new order?

Can the desired alterations be accomplished with less violence?

Aldo Leopold*

1. Explain how the author's statements apply to the flow of energy in the environment.

_____

_____

_____

_____

_____

2. Given your new understanding of how ecosystems work, do you think land-based ecosystems as we know them will "adjust . . . to the new order"? Explain your answer.

_____

_____

_____

_____

_____

_____

_____

_____

READING COMPREHENSION AND ANALYSIS II

Read the following passage, and answer the questions that follow.

The soil exists in a state of constant change, taking part in cycles that have no beginning and no end. New materials are constantly being contributed as rocks disintegrate, as organic matter decays, and as nitrogen and other gases are brought down in rain from the skies. At the same time other materials are being taken away, borrowed for temporary use by living creatures. Subtle and vastly important chemical changes are constantly in progress, converting elements derived from air and water into forms suitable for use by plants. In all these changes living organisms are active agents.

Rachel Carson*

1. How can the soil exist "in a state of constant change" when it continues to play the same role in the cycling of materials?

_____

_____

_____

_____

_____

_____

_____

2. What does the author mean by "cycles that have no beginning and no end"? Explain your answer.

_____

_____

_____

_____

_____

_____

_____

CHAPTER REVIEW

KINDS OF ECOSYSTEMS

Matching   Match each example in the left column with the appropriate term from the right column.

_____ **1.** desert

_____ **2.** tundra

_____ **3.** forest

_____ **4.** grassland

_____ **5.** chaparral

a. cold and dry

b. fertile soils

c. Mediterranean climate

d. canopy vegetation

e. hot and dry

Concept Mapping

6. Complete the climatograms below based on what you have learned about the climate of these different biomes. Vancouver is located in a temperate rain forest, which is characterized by cool, moist winters. The city receives about 130 cm of precipitation per year, a summer high of 18°C, and a winter low of 2°C. Buffalo's climate is typical for a northern deciduous forest. Use a line for temperature and fill in the bar graph for precipitation. It is not important for your answers to be exact but for you to demonstrate that you understand how these climatograms would *generally* look.

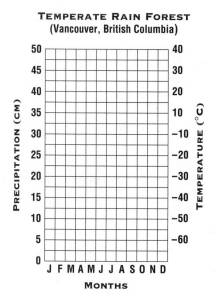

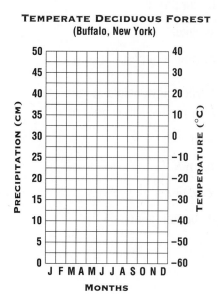

CHAPTER 4 REVIEW, CONTINUED

Multiple Choice   In the space provided, write the letter of the word or statement that best answers the question or completes the sentence.

_____   1. Thin soil, high temperatures, and high rainfall represent a

   a. tropical rain forest.

   b. temperate rain forest.

   c. desert.

   d. grassland.

_____   2. Birds migrating in winter, coniferous plants, and cold temperatures represent a

   a. South Pole.

   b. taiga.

   c. temperate forest.

   d. chaparral.

_____   3. Eutrophication, littoral zone, and zooplankton represent a

   a. grassland.

   b. coral reef.

   c. lake.

   d. photosynthesis.

_____   4. Harbor, phytoplankton, and high productivity represent a

   a. marsh.

   b. river.

   c. estuary.

   d. benthic zone.

_____   5. The North and South Poles can both be considered

   a. terrestrial (land-based) ecosystems.

   b. part of the taiga.

   c. marine ecosystems.

   d. home to penguins.

_____   6. All of the following are examples of freshwater ecosystems EXCEPT

   a. swamps.

   b. estuaries.

   c. marshes.

   d. the Everglades.

_____   7. Productive ecosystems include

   a. estuaries and rain forests.

   b. tundra and savanna.

   c. taiga and desert.

   d. coral reefs and lakes.

_____   8. Factors that influence which plants grow where include

   a. longitude.

   b. climate.

   c. biome maps.

   d. None of these answers are correct.

CHAPTER 4 REVIEW, CONTINUED

Short Answer   Write the answers to the following questions in the spaces provided.

1. Explain how the soil of the tropical rain forest can support the most plant species of any biome yet contain so few nutrients.

_____

_____

_____

_____

_____

_____

2. Describe the role of fire in chaparral and grassland biomes.

_____

_____

_____

_____

_____

_____

3. Describe the difference between the benthic and the littoral zones, and include the organisms you would expect to find in each.

_____

_____

_____

_____

_____

_____

CRITICAL THINKING WORKSHEET

ANALOGIES

Mark the letter of the pair of terms that best completes the analogy shown. An analogy is a relationship between two pairs of words or phrases written as a:b::c:d. The symbol : is read *is to,* and the symbol :: is read *as.*

Example
keyboard : type ::

_____ **a.** plane : land

_____ **b.** dog : eat

√ **c.** scissors : cut

_____ **d.** rock : hard

_____ **e.** grass : green

1. Shallow : deep ::

_____ **a.** tundra : taiga

_____ **b.** littoral : benthic

_____ **c.** ecosystem : biome

_____ **d.** teacher : student

_____ **e.** marsh : swamp

2. Hibernation : cold ::

_____ **a.** ocean : estuary

_____ **b.** estivation : hot

_____ **c.** wet : freezing

_____ **d.** dry : desert

_____ **e.** fur : winter

3. Bromeliads : trees ::

_____ **a.** plants : soil

_____ **b.** fish : water

_____ **c.** fleas : dogs

_____ **d.** water : land

_____ **e.** sponges : coral reef

4. Trees : canopy ::

_____ **a.** bushes : understory

_____ **b.** weeds : bushes

_____ **c.** saplings : trees

_____ **d.** birds : flock

_____ **e.** roots : plants

5. Polar bear : tundra ::

_____ **a.** cactus : desert

_____ **b.** moose : taiga

_____ **c.** desert : cactus

_____ **d.** bison : savanna

_____ **e.** lobster : wetland

6. Phytoplankton : zooplankton ::

_____ **a.** algae : lichens

_____ **b.** reeds : fish

_____ **c.** little fish : big fish

_____ **d.** coral : shrimp

_____ **e.** plants : animals

7. Salty : brackish ::

_____ **a.** swamp : river

_____ **b.** hot : warm

_____ **c.** thermal : heat

_____ **d.** saline : salt

_____ **e.** cool : cold

CRITICAL THINKING WORKSHEET

INTERPRETING DATA

Examine the following data, and answer the questions that follow.

Plant #1: broad leaves, leaves turn yellow in autumn, tall
Plant #2: waxy coating, spines, long and shallow root system
Plant #3: needlelike leaves, pyramid shape, likes acidic soil

1. What is plant #1, which biome is it from, and how is it adapted to that biome?

_____

_____

_____

_____

_____

2. What is plant #2, which biome is it from, and how is it adapted to that biome?

_____

_____

_____

_____

_____

_____

3. What is plant #3, which biome is it from, and how is it adapted to that biome?

_____

_____

_____

_____

_____

_____

_____

CRITICAL THINKING WORKSHEET

AGREE OR DISAGREE

Agree or disagree with the following statements, and support your answer.

1. Many types of ecosystems can exist within a given biome.

_____

_____

_____

_____

_____

_____

2. Humankind's conversion of grasslands to croplands was necessary.

_____

_____

_____

_____

_____

_____

3. Coral reefs are the marine equivalent of tropical rain forests.

_____

_____

_____

_____

_____

_____

CRITICAL THINKING WORKSHEET

REFINING CONCEPTS

The statements below challenge you to refine your understanding of concepts covered in the chapter. Think carefully, and answer the questions that follow.

1. Recommend a strategy for incorporating sustainable human activity into a biome.

   _____

   _____

   _____

   _____

   _____

   _____

2. Is artificial eutrophication undesirable considering that eutrophication occurs naturally anyway? Defend your answer.

   _____

   _____

   _____

   _____

   _____

   _____

3. What would happen if a conifer from the taiga was planted in a tropical rain forest? (Consider roots, growth, and the above ground portions in your answer.)

   _____

   _____

   _____

   _____

   _____

   _____

° C R I T I C A L   T H I N K I N G   W O R K S H E E T   **CHAPTER 4**

READING COMPREHENSION AND ANALYSIS I

Read the following passage and answer the questions that follow.

As in deserts, a limiting physical factor rules these lands (the tundra), but it is heat rather than water that is in short supply in terms of biological functioning. Precipitation is also low, but water as such is not limiting because of the low evaporation rate. Thus, we might think of the tundra as an arctic desert, but it can best be described as a wet arctic grassland or a cold marsh that is frozen for a portion of the year.

Eugene Odum*

1. In your own words, what is the author saying?

_____

_____

_____

_____

2. What does the author mean by limiting factors?

_____

_____

_____

_____

3. Does the author blur the distinction between different biomes (tundra, grasslands, marshes), or does he simply use them to draw an analogy? Defend your answer.

_____

_____

_____

_____

_____

*From *Ecology: A Bridge Between Science and Society* by Eugene P. Odum. Copyright © 1997 by **Sinauer Associates, Inc.** Reprinted by permission of the publisher.

Name _____ Class _____ Date _____

READING COMPREHENSION AND ANALYSIS II

Read the following passage, and answer the questions that follow.

The house knows the sound of El Rio Grande, river that for centuries wandered through this Chihuahua desert, largest desert in North America, old ocean bed where millions of years ago, land emerged from water, mountains rise. Oceans became seas, seas dried to lakes, and lakes evaporated into basins and playas. Water creatures—oysters, clams, coral—hardened in the sea of sand, wordless geological history.

Pat Mora*

1. What point is the author making about the permanence of biomes?

_____

_____

_____

_____

2. Should this point have an effect on our efforts to preserve today's biomes? Justify your response.

_____

_____

_____

_____

_____

3. What effect do you think geological history—the changes of the Earth's surface over time— has on the rise and fall of different biomes?

_____

_____

_____

_____

_____

*From *House of Houses* by Pat Mora. Copyright © 1997 by Pat Mora. Reprinted by permission of **Beacon Press.**

CHAPTER REVIEW

WATER

Matching   Match each example in the left column with the appropriate term from the right column.

_____   **1.** desalinization

_____   **2.** aquifer

_____   **3.** land-based

_____   **4.** bacteria

_____   **5.** multiple sources

_____   **6.** native plants

a. pathogen

b. recharge zone

c. nonpoint pollution

d. ocean pollution

e. reverse osmosis

f. water conservation

True/False   Decide whether the following statements are true or false, and place a T or F in the space to the left of each statement.

_____   **1.** Approximately 0.2 percent of all water on Earth is groundwater.

_____   **2.** Heat energy can be a form of water pollution.

_____   **3.** Nonpoint pollution is a special form of point pollution.

_____   **4.** Polluting in a recharge zone could contaminate an aquifer.

_____   **5.** Biological magnification and artificial eutrophication can both result from water pollution.

_____   **6.** Dams affect ecosystems both upstream and downstream.

_____   **7.** Most of the oil polluting the oceans comes from major oil spills.

_____   **8.** Estuaries bear the brunt of the effects of ocean pollution.

_____   **9.** The rate of groundwater recharge affects the time it takes to decontaminate aquifers.

_____   **10.** In wastewater treatment plants, sedimentation tanks provide
the necessary environment for the aerobic decomposition of sewage.

CHAPTER 5 REVIEW, CONTINUED

Short Answer Write the answers to the following questions in the spaces provided.

1. Explain the relationship between an aquifer and its recharge zone.

_____

_____

_____

_____

_____

2. Define thermal pollution, and provide an example.

_____

_____

_____

_____

_____

3. Compare and contrast point and nonpoint pollution.

_____

_____

_____

_____

_____

4. Explain why only a small portion of the Earth's water is available for human use.

_____

_____

_____

_____

_____

CRITICAL THINKING WORKSHEET

ANALOGIES

Mark the letter of the pair of terms that best completes the analogy shown. An analogy is a relationship between two pairs of words or phrases written as a:b::c:d. The symbol : is read *is to,* and the symbol :: is read *as.*

Example
Keyboard : type ::

_____ **a.** plane : land

_____ **b.** dog : eat

___√___ **c.** scissors : cut

_____ **d.** rock : hard

_____ **e.** grass : green

1. Dam : reservoir ::

_____ **a.** wreck : traffic jam

_____ **b.** braces : teeth

_____ **c.** river : dam

_____ **d.** groundwater : aquifer

_____ **e.** heat : heat wave

2. Artificial eutrophication : fertilizer ::

_____ **a.** pollution : accumulation

_____ **b.** point pollution : nonpoint pollution

_____ **c.** biological magnification : pollution

_____ **d.** evolution : selection

_____ **e.** salinization : salt

3. Water : watershed ::

_____ **a.** food : stomach

_____ **b.** groundwater : aquifer

_____ **c.** blood : arm

_____ **d.** river : lake

_____ **e.** iron : blood

4. Pathogen : *Escherichia coli* ::

_____ **a.** organism : biotic

_____ **b.** pathogen : feces

_____ **c.** parasite : flea

_____ **d.** fish : organism

_____ **e.** bacteria : disease

5. Nonpoint pollution : multiple sources ::

_____ **a.** pesticides : salt

_____ **b.** point pollution : single source

_____ **c.** biotic factors : pathogen

_____ **d.** aquifer : recharge zone

_____ **e.** eutrophication : succession

6. Desalination : salt ::

_____ **a.** pumping : water

_____ **b.** pumping : aquifer

_____ **c.** pollution : decontamination

_____ **d.** dehydration : water

_____ **e.** eutrophication : sewage

7. Water pollutant : pathogen ::

_____ **a.** thermal pollution : physical

_____ **b.** vegetable : broccoli

_____ **c.** point pollution : street oil

_____ **d.** septic tank : point pollution

_____ **e.** pathogens : DDT

CRITICAL THINKING WORKSHEET

CHAPTER
5

AGREE OR DISAGREE

Agree or disagree with the following statements, and support your answer.

1. A river is greater than the sum of the streams that feed it.

_____

_____

_____

_____

_____

_____

2. Ocean pollution is more likely to affect a large sea animal, like a whale, than a small shoreline animal, like a crab.

_____

_____

_____

_____

_____

_____

3. Pumping and diverting water affects the water cycle in predictable ways.

_____

_____

_____

_____

_____

_____

_____

CRITICAL THINKING WORKSHEET CHAPTER

THINKING SCIENTIFICALLY

Read the following scenario, and answer the questions that follow.

On your way to school you notice that the water in a previously clear local stream now appears brown. You have heard that a new paper-processing plant recently opened upstream, and you wonder if there might be a connection.

1. Form a hypothesis based on your observations.

   _____

   _____

   _____

2. How would you test your hypothesis and evaluate your results?

   _____

   _____

   _____

   _____

   _____

   _____

   _____

3. Assume that your data seem to contradict your hypothesis. What can you conclude?

   _____

   _____

4. How might you construct a new hypothesis based on the information you collected?

   _____

   _____

   _____

   _____

Name _____ Class _____ Date _____

READING COMPREHENSION AND ANALYSIS

Read the following passage, and answer the questions that follow.

Everything depends on the manipulation of water—on capturing it behind dams, storing it, and rerouting it in concrete rivers over distances of hundreds of miles. Were it not for a century and a half of messianic effort toward that end, the West as we know it would not exist.

Marc Reisner*

1. The word *messianic* is the adjective form of the word *messiah*. A messiah is an expected savior or liberator of a people or country. Why do you think the author used this word to describe the effort to manipulate water in the West?

   _____

   _____

2. In your own words, what is the author saying?

   _____

   _____

   _____

3. What do you think is the "everything" that the author refers to in the first line?

   _____

   _____

   _____

4. Does your understanding of water supply issues in the western United States support the author's claim? Explain.

   _____

   _____

   _____

   _____

CHAPTER REVIEW

AIR

Matching Match each example in the left column with the appropriate term from the right column.

_____ 1. primary pollutant

_____ 2. secondary pollutant

_____ 3. indoor air pollution

_____ 4. pollution control

_____ 5. acid precipitation

a. ozone

b. scrubber

c. radon gas

d. nitrogen oxides

e. chemical weathering

Concept Mapping

6. Complete the diagram below by drawing lines between categories in adjacent columns that relate to each other.

Locations	Sources of pollution	Possible outputs
		Nitrogen oxides
Indoors	Natural	Bacteria
		Pollen
		Asbestos
		Sulfur dioxide
		Radon
Outdoors	Human activities	Fungi
		Dust
		Formaldehyde
		Carbon dioxide

CHAPTER 6 REVIEW, CONTINUED

Multiple Choice   In the space provided, write the letter of the word or statement that best answers the question, completes the sentence, or ties the words together.

_____ 1. Carbon monoxide, sulfur dioxide, particulate matter

    a. secondary pollutants

    b. primary pollutants

    c. thermal inversions

    d. primary sources

_____ 2. Tightly-sealed, asbestos, poor ventilation

    a. thermal inversion

    b. radon gas

    c. sick-building syndrome

    d. acid shock

_____ 3. Catalytic converters, scrubbers, electrostatic precipitators

    a. sick-building syndrome

    b. VOCs

    c. air pollution

    d. Clean Air Act

_____ 4. Warm air, cool air, pollutants

    a. primary sources

    b. industrial processes

    c. smog

    d. thermal inversion

_____ 5. Dead trees, lime, sulfuric acid

    a. acid precipitation

    b. indoor air pollution

    c. scrubber

    d. toxic chemicals

_____ 6. Sugar cane, hydrocarbons, corn, cars

    a. organic compounds

    b. alternative fuels

    c. fossil fuels

    d. ethanol

_____ 7. Paint thinner, gasoline, cars

    a. particulate matter

    b. VOCs

    c. Both a and b are correct.

    d. industrial effects

_____ 8. Uranium, houses, rocks

    a. sick-building syndrome

    b. indoor air pollution

    c. radon

    d. primary sources

CHAPTER 6 REVIEW, CONTINUED

Short Answer   Write the answers to the following questions in the spaces provided.

1. Name and describe three effects of air pollution on health.

   _____

   _____

   _____

   _____

   _____

   _____

   _____

2. How is mass transit in cities a possible solution to urban air pollution?

   _____

   _____

   _____

   _____

   _____

   _____

   _____

   _____

3. Describe two atmospheric conditions that make air pollution worse.

   _____

   _____

   _____

   _____

   _____

   _____

   _____

   _____

Name _____ Class _____ Date _____

INTERPRETING OBSERVATIONS

Read the following scenario, and answer the questions that follow.

Lake Sulfox seems to be having some problems with its fish population. Commercial fisherman are claiming that their catches have declined, and they are blaming the decline on the supposed acidification of the lake by a local coal-fired power plant. The Lake Sulfox Advisory Board has the following data on file. Assume that the size of the fish harvest is a good indicator of the size of the fish population.

Annual Fish Harvest (metric tons)

1991	1992	1993	1994	1995	1996
7500	6924	6322	5412	5503	5113

Mean Sulfate Levels (ppm)

1991	1992	1993	1994	1995	1996
41.07	51.34	54.89	57.46	58.76	59.65

1. What is the relationship between the size of the fish harvest and the sulfate levels in the lake?

    _____

    _____

    _____

2. Do the data prove that acidification of the lake by sulfates is responsible for the decline in the lake's fish population? Provide at least two reasons to support your viewpoint.

    _____

    _____

    _____

    _____

    _____

    _____

    _____

    _____

CRITICAL THINKING WORKSHEET

CHAPTER

6

REFINING CONCEPTS

The statements below challenge you to refine your understanding of concepts covered in the chapter. Think carefully, and answer the questions that follow.

1. Imagine that you are the city manager, and the EPA has given your city a citation for high levels of oxides in the air. What steps would you take to reduce the levels of oxides in the air?

    _____

    _____

    _____

    _____

    _____

2. Where might using grain as an alternative fuel source not be a good idea? Explain.

    _____

    _____

    _____

    _____

    _____

3. Manufacturing is often blamed for producing air pollution. However, according to economic theory, it is neither technologically feasible nor economically efficient to completely eliminate pollution. What do you think? Explain your reasoning.

    _____

    _____

    _____

    _____

    _____

    _____

    _____

44

CRITICAL THINKING WORKSHEET

READING COMPREHENSION AND ANALYSIS

Read the following passage, and answer the questions that follow.

If the wind can carry seeds and spores and the fragrance of hay, it can also carry man-made molecules, the gases and compounds exhaled by industrial activities. We have, in fact, depended on it to do so, building higher smokestacks in hopes that the air currents farther up will carry our smoke farther away, out of sight, out of mind, eliminating them as a local problem, at least.

Stephanie Mills*

1. Do you agree with the author's idea that air pollution is often an "out of sight, out of mind" problem? Explain your answer.

   _____

   _____

   _____

   _____

2. How do you think limiting the height of smokestacks would affect the air pollution problem?

   _____

   _____

   _____

   _____

3. Many environmentalists have proposed selling a limited number of permits to facilities that release pollutants into the air. In theory, any person or group could buy the permits. How could this help solve the air pollution problem?

   _____

   _____

   _____

   _____

*From *In Praise of Nature* by Stephanie Mills. Copyright © 1990 by Island Press. Reprinted by permission of *Alexander Hoyt Associates.*

CHAPTER REVIEW

ATMOSPHERE AND CLIMATE

Matching   Match each example in the left column with the appropriate term from the right column.

_____ **1.** increased snow cover

_____ **2.** CFCs in atmosphere

_____ **3.** ocean warming

_____ **4.** burning fossil fuels

_____ **5.** low-angle sunlight

_____ **6.** high levels of ozone in upper atmosphere

_____ **7.** increased carbon dioxide in atmosphere

_____ **8.** high UV radiation at Earth's surface

a. less UV radiation reaches Earth's surface

b. more CO_2 in atmosphere

c. ozone destruction

d. cooler temperatures

e. more sunlight reflected from Earth

f. heat is trapped near surface

g. increased DNA damage

h. more water vapor in atmosphere

Concept Mapping

9. The climate at the equator is wet and rainy but very dry at latitudes 30 degrees north and south of the equator. Climate is the average weather pattern of a region over time. Complete the concept map below to illustrate the processes responsible for a region's climate. Use the following terms: *air circulation, precipitation, ocean currents, latitude, local geography, Earth's rotation, solar energy received,* and *winds.*

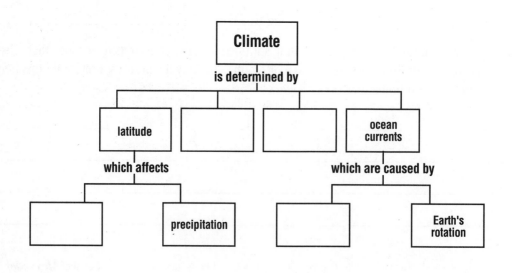

Name _____ Class _____ Date _____

Multiple Choice In the space provided, write the letter of the statement that best answers the question or completes the sentence.

_____ 1. The weather we experience occurs in the

    a. troposphere.

    b. stratosphere.

    c. ozone layer.

    d. exosphere.

_____ 2. Rain is common whenever

    a. cold, moist air rises.

    b. warm, moist air rises.

    c. warm, dry air sinks.

    d. cold, dry air sinks.

_____ 3. Earth's atmosphere is the only place in the solar system with large amounts of

    a. water vapor.

    b. methane.

    c. oxygen.

    d. weather.

_____ 4. Cloud cover makes the ground-level temperature

    a. cooler by day.

    b. cooler by night.

    c. warmer by night.

    d. Both a and c are correct.

_____ 5. The gas most responsible for the greenhouse effect is

    a. nitrous oxide.

    b. methane.

    c. oxygen.

    d. water vapor.

_____ 6. Which of the following reduce CO_2 in the atmosphere?

    a. phytoplankton

    b. tropical rain forests

    c. oceans

    d. All of the answers are correct.

_____ 7. In the summer, the northern hemisphere gets sunlight

    a. obliquely for long days.

    b. slanting for short days.

    c. more directly for long days.

    d. less directly for short days.

_____ 8. Ozone in the stratosphere

    a. causes skin cancer.

    b. prevents DNA repair.

    c. absorbs UV light.

    d. destroys CFCs.

CHAPTER 7 REVIEW, CONTINUED

Short Answer   Write the answers to the following questions in the spaces provided.

1. Human activities currently release 7 billion tons of carbon dioxide into the atmosphere each year. What is the source of this excess carbon dioxide?

    _____

    _____

    _____

    _____

2. What are the most important steps industrialized countries could take to help reduce levels of atmospheric carbon dioxide?

    _____

    _____

    _____

    _____

    _____

    _____

    _____

3. Why are some developed countries willing to pay to help developing countries find substitutes for CFCs?

    _____

    _____

    _____

    _____

4. Why is it important for politicians to understand scientific information on the greenhouse effect and ozone holes?

    _____

    _____

    _____

    _____

CRITICAL THINKING WORKSHEET

INTERPRETING EVIDENCE

Read the following passage, and answer the questions that follow.

Alaska is thawing, as is much of Canada and northern Russia. Hundreds of glaciers are retreating. The warmer atmosphere has produced more snow to feed the glacier, but longer, warmer summers melt them faster than the heavier snows can build them. The region's permafrost is thawing in the interior, and pockets of underground ice trapped in the frost are melting too. Forests are drowning as the ground sinks and water floods their roots. Trees, weakened by climate-related stresses, are killed by spruce bark beetles whose population has exploded. Average global temperatures have increased by 1°F over the last century. But in Alaska, Siberia, and northwest Canada, average temperatures have increased as much as 5°F over the last 30 years. Warming is more pronounced in winter. Mainstream scientists predict that Alaska is expected to warm twice as much as the global average.

Anne Gregory*

1. After warming starts, is a snow-ice region likely to show a sharper temperature rise than a region without snow-ice? Explain your answer.

_____

_____

_____

_____

_____

_____

2. Do you agree with the mainstream scientists' prediction about Alaska? Justify your response.

_____

_____

_____

_____

_____

_____

_____

_____

Name _____ Class _____ Date _____

AGREE OR DISAGREE

Agree or disagree with the following statements, and support your answer.

1. Industrial countries should assist tropical rain forest countries so those countries can afford to leave their forests intact.

   _____

   _____

   _____

   _____

   _____

   _____

   _____

   _____

2. The correlation between carbon dioxide levels in the atmosphere and world temperatures for the past 160,000 years proves that higher carbon dioxide levels cause global warming.

   _____

   _____

   _____

   _____

   _____

   _____

3. Developing countries should not participate in treaties that set allowable levels of greenhouse emissions in developed countries.

   _____

   _____

   _____

   _____

CRITICAL THINKING WORKSHEET

REFINING CONCEPTS

The statements below challenge you to refine your understanding of concepts covered in the chapter. Think carefully, and answer the questions that follow.

1. Some scientists predict that global warming will cause major ocean currents to shut down. The Gulf Stream moves warm water toward northern latitudes, and the South Polar Conveyor Belt moves cold water toward the Equator. How might an ocean current shutdown affect the climate?

   _____

   _____

   _____

   _____

2. A catalyst speeds up a process but is not changed itself. CFCs are known to release catalysts that break down the ozone layer. How does this process work?

   _____

   _____

   _____

   _____

   _____

3. The carbon in fossil fuels was in the atmosphere long ago. Why is it now a problem that we burn those fossil fuels and put the carbon back into the atmosphere?

   _____

   _____

   _____

   _____

   _____

Name _____ Class _____ Date _____

READING COMPREHENSION AND ANALYSIS

Read the following passage, and answer the questions that follow.

So, when the earth's atmosphere acquired its oxygen from the photosynthetic activity of green plants, the planet also acquired a high-altitude blanket of ozone. Until then the earth's surface was bathed in intense ultraviolet radiation, which was, in fact, the energy source that converted the early earth's blanket of methane, water, and ammonia into the soup of organic compounds in which the first living things originated. However, ultraviolet radiation is very damaging to the delicate balance of chemical reactions in living cells, and it is likely that the first living things survived only by growing under a layer of water sufficiently thick to protect them from the ultraviolet radiation that reached the earth's surface.

Barry Commoner*

1. How might the decrease in UV radiation reaching Earth have affected the evolution of life-forms?

   _____

   _____

   _____

   _____

2. Read the last sentence of the passage again. How does the ozone layer differ from the water that shielded early organisms from UV light?

   _____

   _____

   _____

3. Do you think the human species could survive without the ozone layer? Explain your answer.

   _____

   _____

   _____

   _____

---

*From *The Closing Circle: Nature, Man, and Technology* by Barry Commoner. Copyright © 1971 by Barry Commoner. Reprinted by permission of *Alfred A. Knopf, Inc.*

CHAPTER REVIEW

LAND

Matching   Match each example in the left column with the appropriate term from the right column.

_____ **1.** infrastructure

_____ **2.** urbanization

_____ **3.** deforestation

_____ **4.** mineral resource

_____ **5.** wilderness

_____ **6.** mining

_____ **7.** public land

a. multiple use

b. protected land

c. reclamation

d. bridges

e. sulfur

f. development

g. clear-cutting

Ranking   Human activity changes the land. Rank the following types of land use according to how much the landscape has been altered by humans. Rank the least changed category 1 and the most changed 7.

Land Use Type	Landscape Change Rank (1–7)
8. corn farm	_____
9. national park	_____
10. rangeland	_____
11. suburban shopping mall	_____
12. abandoned wheat field	_____
13. downtown business district	_____
14. managed forest	_____

CHAPTER 8 REVIEW, CONTINUED

Multiple Choice   In the space provided, write the letter of the word or statement that best answers the question or completes the sentence.

_____ 1. Reclamation reduces the ultimate impact of

    a. mining.

    b. crime.

    c. overgrazing.

    d. air pollution.

_____ 2. All of the following are found in wilderness except

    a. habitat.

    b. camping.

    c. suburbs.

    d. fishing.

_____ 3. A modern consequence of our growing urban areas is

    a. lower crime rate.

    b. suburban sprawl.

    c. infrastructure.

    d. energy conservation.

_____ 4. Open pit mining creates

    a. ore.

    b. holes.

    c. strips.

    d. mineral resources.

_____ 5. The "wedges and corridors" plan is an example of

    a. architecture.

    b. urban crisis.

    c. highway construction.

    d. land-use planning.

_____ 6. Tree-harvesting methods include

    a. selective cutting.

    b. reforestation.

    c. clear-cutting.

    d. Both a and c are correct.

_____ 7. A fire station is an example of

    a. infrastructure.

    b. suburbanization.

    c. land-use planning.

    d. renovation.

_____ 8. Rangeland may suffer from

    a. overgrazing.

    b. desertification.

    c. poor management.

    d. All of the above answers are correct.

Name _____ Class _____ Date _____

Short Answer   Write the answers to the following questions in the spaces below.

1. Explain the advantages and disadvantages of clear-cutting as a method of harvesting timber.

_____

_____

_____

_____

_____

_____

2. Explain the role of inadequate infrastructure in the urban crisis.

_____

_____

_____

_____

_____

_____

3. Describe two methods for managing rangeland.

_____

_____

_____

_____

_____

_____

_____

_____

CHAPTER 8 REVIEW, CONTINUED

Understanding Vocabulary   For each pair of terms explain the difference in their meanings.

1. *Suburban sprawl* and *urban crisis*

_____

_____

_____

_____

_____

2. *Clear-cutting* and *selective cutting*

_____

_____

_____

_____

_____

3. *Deforestation* and *desertification*

_____

_____

_____

_____

_____

4. *Reforestation* and *reclamation*

_____

_____

_____

_____

_____

_____

CRITICAL THINKING WORKSHEET

AGREE OR DISAGREE

Agree or disagree with the following statements, and support your answer.

1. It is more economically advantageous in the long run to protect or preserve open space.

_____

_____

_____

_____

_____

_____

_____

_____

2. Mining companies should restore land to the same successional stage it was in when they mined it. Figure 3-16, on page 67, illustrates different successional stages of a forest.

_____

_____

_____

_____

_____

_____

3. The lack of adequate subway systems in most urban areas of the United States reflects poor land-use planning.

_____

_____

_____

_____

_____

CRITICAL THINKING WORKSHEET CHAPTER 8

REFINING CONCEPTS

The statements below challenge you to refine your understanding of concepts covered in the chapter. Think carefully, and answer the questions that follow.

1. Mining dramatically alters the form and function of land. What ecological challenges do you think a mining reclamation team might face?

   _____

   _____

   _____

   _____

   _____

2. If you were a land-use planner, what guidelines would you use to locate a new mall on the outskirts of your community?

   _____

   _____

   _____

   _____

   _____

3. What issues in your community or in the nearest city could be considered part of the urban crisis? Discuss these issues.

   _____

   _____

   _____

   _____

   _____

   _____

   _____

CRITICAL THINKING WORKSHEET

CHAPTER

8

READING COMPREHENSION AND ANALYSIS I

Read the following passage, and answer the questions that follow.

The bare vastness of the Hopi landscape emphasizes the visual impact of every plant, every rock, every arroyo. Nothing is overlooked or taken for granted. Each ant, each lizard, each lark is imbued with great value simply because the creature is there, simply because the creature is alive in a place where any life at all is precious. . . . So little lies between you and the earth. One look and you know that simply to survive is a great triumph, that every possible resource is needed, every possible ally—even the most humble insect or reptile. You realize you will be speaking with all of them if you intend to last out the year.

Leslie Marmon Silko*

1. What do you think is the author's overall message in this passage?

   _____

   _____

   _____

   _____

2. What sort of attitude does the author have toward the land and its resources?

   _____

   _____

   _____

3. How do we "speak" to the many land-based resources that we use?

   _____

   _____

   _____

   _____

   _____

   _____

Name _____ Class _____ Date _____

CHAPTER
8

READING COMPREHENSION AND ANALYSIS II

Read the following passage, and answer the questions that follow.

There is yet no ethic dealing with man's relation to land and to the animals and plants which grow upon it . . . land is still property. The land-relation is still strictly economic, entailing privileges but not obligations.

Aldo Leopold*

1. What are the "privileges and obligations" that the author is referring to?

_____

_____

_____

_____

_____

_____

2. Explain how we can use methods outlined in the chapter to fulfill our "obligations."

_____

_____

_____

_____

3. Who do you think is responsible for instituting a land ethic? Explain your answer.

_____

_____

_____

_____

*From *A Sand County Almanac* by Aldo Leopold. Copyright © 1949, 1977 by **Oxford University Press, Inc.** Reprinted by permission of the publisher.

CHAPTER REVIEW

FOOD

Matching   Match each example in the left column with the appropriate term from the right column.

_____ **1.** clearing forests

_____ **2.** overuse of land

_____ **3.** fertile soil

_____ **4.** high pesticide use

_____ **5.** population pressure

_____ **6.** no-till farming

_____ **7.** irrigation and evaporation

a. less erosion

b. resistance to pesticides

c. salinization

d. malnutrition or starvation

e. desertification

f. action of living organisms

g. loss of topsoil

Concept Mapping

8. Complete the unfinished diagram below to illustrate the connections between the different components.

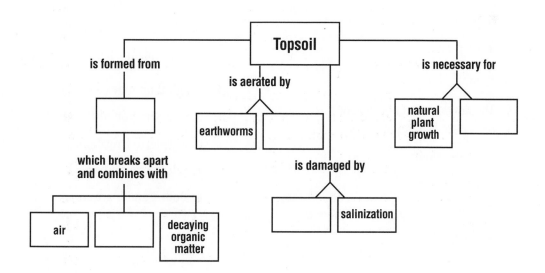

CHAPTER 9 REVIEW, CONTINUED

Multiple Choice   In the space provided, write the letter of the word or statement that best answers the question or completes the sentence.

_____ 1. The biggest loss of arable land is caused by

    a. erosion due to floods and droughts.

    b. salinization and desertification.

    c. use of farmland for urban development.

    d. climatic changes.

_____ 2. The living organisms in fertile soil are found in

    a. the surface litter and topsoil.

    b. the leaching zone.

    c. the subsoil.

    d. the bedrock.

_____ 3. Organic farmers use all of the following EXCEPT

    a. compost and animal manure.

    b. pesticides such as malathion.

    c. crop rotation.

    d. cover crops.

_____ 4. The development of pesticide resistance is an example of

    a. malnutrition.

    b. persistence.

    c. pest control.

    d. evolution.

_____ 5. The green revolution depended on

    a. new biodegradable pesticides.

    b. high-yielding grain varieties.

    c. clearing forest for crop land.

    d. powerful fertilizers.

_____ 6. Erosion is a danger whenever the soil is

    a. bare and exposed to wind and rain.

    b. plowed along the contour of the land.

    c. covered with grass.

    d. planted to forest.

_____ 7. Biological pest control aims to do all the following EXCEPT

    a. maintain tolerable pest levels.

    b. reduce all insects to low levels.

    c. leave non-pest species unharmed.

    d. boost plants' natural defenses.

_____ 8. Salinization may be caused by

    a. a rise in groundwater levels.

    b. long-term irrigation.

    c. salt-tolerant crops.

    d. Both a and b are correct.

CHAPTER 9 REVIEW, CONTINUED

Short Answer   Write the answers to the following questions in the spaces provided.

1. Explain how biological insect control can kill only the target pest while chemical insecticides kill many different kinds of insects.

_____

_____

_____

_____

_____

2. Explain why heavy use of chemical insecticides means that we have to keep developing new ones to use.

_____

_____

_____

_____

_____

3. Explain how political problems can be more important than agricultural yields in determining whether people go hungry.

_____

_____

_____

_____

_____

_____

4. Describe two farming practices that can help reduce erosion from water.

_____

_____

_____

_____

_____

_____

Name _____ Class _____ Date _____

AGREE OR DISAGREE

Agree or disagree with the following statements, and support your answer.

1. If we develop salt-tolerant plants, we may be able to use ocean water for crop irrigation.

_____

_____

_____

_____

_____

2. The reason the world's deserts are expanding is that the greenhouse effect is warming the climate.

_____

_____

_____

_____

_____

3. It would be better to apply genetic engineering to develop resistant crops than to rely on chemical pesticides.

_____

_____

_____

_____

_____

_____

4. The green revolution made it possible for subsistence farmers to grow more and produce a surplus to sell.

_____

_____

_____

_____

CRITICAL THINKING WORKSHEET

REFINING CONCEPTS

The statements below challenge you to refine your understanding of concepts covered in the chapter. Think carefully, and answer the questions that follow.

1. Clearing land for crops and the resulting destruction of habitat is leading to extinction of plants and animals at a high rate. Why do plant breeders and genetic engineers need to have as broad a base of "wild" plants as possible?

_____

_____

_____

_____

_____

_____

2. Organic farming adds material to the litter layer. How does this affect the soil profile?

_____

_____

_____

_____

_____

_____

3. People have been farming for the past 10,000 years. If erosion had been going on at the current rate for all that time, we would have reached bedrock long ago. Why has erosion become a serious problem only recently?

_____

_____

_____

_____

_____

_____

_____

CRITICAL THINKING WORKSHEET

CHAPTER 9

INTERPRETING DATA

Read the following information, and answer the questions that follow.

Rice, corn, millet, and wheat are four of the most widely eaten grains. In the raw, these grains each provide 330 Calories or more of energy per 100 g of grain. However, cooking greatly reduces the available energy, especially in rice and millet. Cooked rice provides only about 110 Cal/100 g, and cooked millet about 120 Cal/100 g.

1. Refer to Figure 9-4, on page 229.

   a. Calculate the amount of grain produced per person per day in 1995.

   b. Assume that the average amount of energy provided by cooked grains is 200 Cal/100 g. Calculate how many Calories/person/day were available from grain produced in 1995 assuming that we ate all of the grain we produced.

   c. Adults require between 2,000 and 3,500 Cal/day, while children usually require less. Assume that the average person requires 1,800 Cal/day to remain healthy. Assuming that we can subsist entirely on grain, calculate whether we produced enough grain to feed the people of the world in 1995. Note that this is an oversimplification because humans need a varied diet to stay healthy.

2. Refer to Figure 9-7, on page 231. What percentage of worldwide arable land was lost between 1985 and 2000?

CRITICAL THINKING WORKSHEET

READING COMPREHENSION AND ANALYSIS I

Read the following passage, and answer the questions that follow.

There are two quite simple reasons why I believe that the world will continue to depend on farming, supplemented in an important way by fishing, for most of its food supply in the decades ahead. The first is that the principal food of the world (eaten directly or through animal products), the grains, is relatively low in cost.…The second reason is the enormous *weight* of food that is produced each year. In order to supply the calories required for life, it appears that there is little possibility of reducing the dry weight involved. U.S. grain production, measured by *weight,* is more than 1.5 times greater than steel and 10 times greater than automobiles. The current low cost of food grains—the major primary or secondary source of calories for all people—combined with the enormous volume or weight involved make it most unlikely that much progress will be made in the next fifty years in replacing agriculture by factories.

D. Gale Johnson*

1. What is the source of carbon in most crops today, and what would probably be the carbon source in factory-made crops?

_____

_____

_____

2. How does the source of energy to make today's crops differ from the possible sources of energy we would need to make crops in factories?

_____

_____

_____

_____

3. The author wrote this article more than thirty years ago. Do you think he was right or wrong?

_____

_____

_____

*From "Food: The World's People Won't Go Hungry" by **D. Gale Johnson** from *Toward the Year 2018*, edited by the Foreign Policy Association. Copyright © 1968 by the Foreign Policy Association. Reprinted by permission of ***the author and the Foreign Policy Association.***

Name _____ Class _____ Date _____

READING COMPREHENSION AND ANALYSIS II

Read the following passage, and answer the questions that follow.

Under primitive agricultural conditions the farmer had few insect problems. These arose with the intensification of agriculture—the devotion of immense acreages to a single crop. Such a system set the stage for explosive increases in specific insect populations. Single-crop farming does not take advantage of the principles by which nature works; it is agriculture as an engineer might conceive it to be.

Rachel Carson*

1. Why would planting large areas with a single type of crop plant "set the stage" for large increases in insect populations?

   _____

   _____

   _____

   _____

   _____

   _____

2. How does single-crop farming, also known as monocropping, "not take advantage of the principles by which nature works?"

   _____

   _____

   _____

   _____

   _____

   _____

   _____

   _____

   _____

CHAPTER REVIEW

CHAPTER
10

BIODIVERSITY

Matching   Match each example in the left column with the appropriate term from the right column.

_____   **1.** a species not native to a particular region

_____   **2.** any species that is likely to become
endangered if they are not protected

_____   **3.** species that are very important to the
functioning of an ecosystem

_____   **4.** any species whose numbers have fallen so
low that it is likely to become extinct in the
near future if not protected immediately

_____   **5.** status of a species when the very last
individual dies

a. keystone species

b. exotic species

c. extinct

d. endangered species

e. threatened species

Now match each definition in the left column with the vocabulary word in the right column.

_____   **6.** Florida panther

_____   **7.** American passenger pigeon

_____   **8.** sea otter

_____   **9.** northern spotted owls

_____   **10.** melaleuca trees

a. keystone species

b. exotic species

c. extinct species

d. endangered species

e. threatened species

CHAPTER 10 REVIEW, CONTINUED

Concept Mapping

11. In each blank circle, write the name of the organism from the list that corresponds to the specific threat in the attached box. Then draw a line from each of the specific threats to the corresponding general threat in the middle of the diagram.

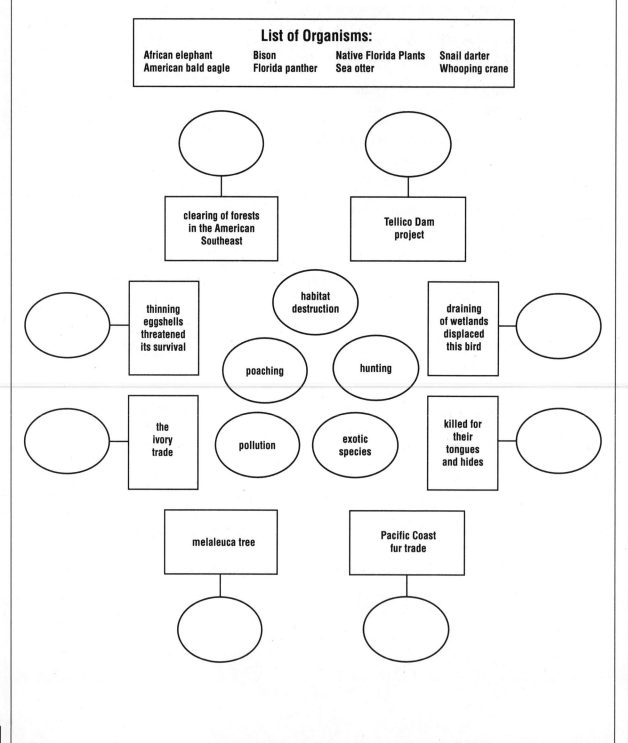

List of Organisms:

African elephant	Bison	Native Florida Plants	Snail darter
American bald eagle	Florida panther	Sea otter	Whooping crane

clearing of forests in the American Southeast

Tellico Dam project

habitat destruction

thinning eggshells threatened its survival

draining of wetlands displaced this bird

poaching

hunting

the ivory trade

pollution

exotic species

killed for their tongues and hides

melaleuca tree

Pacific Coast fur trade

Name _____ Class _____ Date _____

Short Answer Write the answers to the following questions in the spaces provided.

1. Explain why many conservationists now concentrate on protecting entire ecosystems rather than individual species.

_____

_____

_____

_____

_____

2. Explain the difference between an *endangered* species and a *threatened* species.

_____

_____

_____

_____

_____

3. Briefly explain three ways to save individual species.

_____

_____

_____

_____

_____

_____

_____

4. Why do many scientists believe that the Earth is currently experiencing a mass extinction?

_____

_____

_____

_____

_____

_____

71

Name ______________________ Class ______________ Date ______________

CRITICAL THINKING WORKSHEET

CHAPTER 10

ANALYZING ARGUMENTS

The gray wolf is a species that people have maligned in campfire tales and stories such as *Little Red Riding Hood*. Recently, the gray wolf was reintroduced into Yellowstone National Park. Shortly thereafter, a federal judge ruled that the reintroduction program was illegal for various reasons.

1. Why is reintroducing the gray wolf important for the Yellowstone ecosystem?

____

____

____

____

____

____

2. Nearby ranchers claim that the wolves will prey on their livestock. Is it possible to evaluate this claim "before the fact?"

____

____

____

____

____

____

____

3. Is it possible to predict the effect(s) of eliminating a species from an ecosystem? Explain your answer.

____

____

____

____

____

____

____

CRITICAL THINKING WORKSHEET

AGREE OR DISAGREE

Agree or disagree with the following statements, and support your answer.

1. When battles between developers and environmentalists are worked out, usually neither side gets everything they want, but both sides get something. This compromising approach is effective enough to save endangered species.

_____

_____

_____

_____

_____

_____

2. Do you think the Biodiversity Treaty should have been signed?

_____

_____

_____

_____

_____

_____

_____

3. To protect biodiversity worldwide, many conservationists suggest that at least 10 percent of the Earth's land be set aside as protected preserves. This percentage is a good "hard and fast" rule to be used in all cases.

_____

_____

_____

_____

_____

_____

_____

_____

CRITICAL THINKING WORKSHEET

INTERPRETING EVIDENCE

Write the answers to the following questions in the spaces provided.

1. Migratory species, such as many birds, salmon, and the monarch butterfly, can be especially vulnerable to habitat loss. At the same time, they can be instrumental in preserving ecosystems. Explain why both of these statements are true.

   _____
   _____
   _____
   _____
   _____
   _____
   _____

2. What conditions do the Florida panthers need in order to come back from the brink of extinction? Why is this unlikely to occur?

   _____
   _____
   _____
   _____
   _____

3. In the summer of 1998, numerous fires spread through parts of Florida. Explain how importing melaleuca trees in the early 1900s may have contributed to this situation.

   _____
   _____
   _____
   _____
   _____
   _____

CRITICAL THINKING WORKSHEET

READING COMPREHENSION AND ANALYSIS I

Read the following passage, and answer the questions that follow.

The immense diversity of the insects and flowering plants combined is no accident. The two empires are united by intricate symbioses. The insects consume every anatomical part of the plants, while dwelling on them in every nook and cranny. A large fraction of the plant species depend on insects for pollination and reproduction. Ultimately they owe them their very lives, because insects turn the very soil around their roots and decompose dead tissues into the nutrients required for continued growth.

So important are insects and other land-dwelling arthropods that if all were to disappear, humanity probably could not last more than a few months.

Edward O. Wilson*

1. What type of species interaction is the author referring to (hint: see Chapter 2), and what does it imply about the evolution of the Earth's biodiversity?

   _____

   _____

   _____

   _____

2. What is the reasoning behind the author's last statement?

   _____

   _____

   _____

   _____

3. Popular images of biodiversity often focus on large endangered mammals, such as pandas and tigers. Why do you think the actual situation is not as well-known?

   _____

   _____

   _____

   _____

CRITICAL THINKING WORKSHEET

READING COMPREHENSION AND ANALYSIS II

Read the following passage, and answer the questions that follow.

Coffee is traditionally grown under a canopy of shade trees. Because of the structural and floristic complexity of the shade trees, traditional coffee plantations have relatively high biodiversity. However, coffee plantations increasingly are being transformed into industrial plantations with little or no shade.... The way that coffee production evolves in the coming decades is likely to have a tremendous impact on its ability to provide a refuge for tropical biodiversity.

Ivette Perfecto et al.*

1. The author states that traditional coffee plantations have relatively high biodiversity. What does she mean by this statement?

_____

_____

_____

_____

2. The author offers an explanation for the biodiversity of traditional coffee plantations. What does this explanation imply about the biodiversity of tropical rain forests?

_____

_____

_____

_____

3. What point is the author making in her last sentence?

_____

_____

_____

_____

_____

*From "Shade Coffee: A Disappearing Refuge for Biodiversity" by Ivette Perfecto et al. from *BioScience*, vol. 46, no. 8, September 1996, p. 598. Copyright © 1996 by the **American Institute of Biological Sciences.** Reprinted by permission of the publisher.

CHAPTER REVIEW

ENERGY

Matching   Match each example in the left column with the appropriate term from the right column.

_____ **1.** biomass, sun, wind, water

_____ **2.** fission and fusion

_____ **3.** oil, natural gas, and coal

_____ **4.** magnetic fields and rotating turbines

_____ **5.** sugar cane, litter from chicken coops, and wood

a. fossil fuels

b. renewable resources

c. electric generator

d. biomass

e. nuclear energy

Concept Mapping

6. Demonstrate how energy is converted using the diagram below. Draw lines connecting each energy source through the complete process of energy conversion. Each energy source may be converted through more than one process.

Sources	Energy conversions	Possible outputs
Burning biomass	heat(s) water to make steam that spins turbines in an electric generator	to produce electricity
Dams	wind spins turbines in an electric generator	
Hot rocks in the Earth's crust		
Nuclear reactor	water spins turbines in an electric generator	
Sunlight	is converted by means of solar cells	
Waterfalls		to produce heat
Windmills	is converted by means of solar collectors	

REVIEW AND CRITICAL THINKING WORKSHEETS   **CHAPTER 11 • REVIEW**

CHAPTER 11 REVIEW, CONTINUED

Short Answer   Write the answers to the following questions in the spaces provided.

1. Distinguish between renewable and nonrenewable resources. Provide a specific example of each.

_____

_____

_____

_____

_____

2. How is a nuclear power plant similar to a plant that burns fossil fuels? How are they different?

_____

_____

_____

_____

_____

_____

3. Define energy conservation, and give three examples not listed in the book.

_____

_____

_____

_____

_____

4. Is wind energy a form of solar energy? Justify your answer.

_____

_____

_____

_____

Name _____ Class _____ Date _____

ANALOGIES

Mark the letter of the pair of terms that best completes the analogy shown. An analogy is a relationship between two pairs of words or phrases written as a:b::c:d. The symbol : is read *is to,* and the symbol :: is read *as.*

Example
keyboard : type ::

_____ **a.** plane : land

_____ **b.** dog : eat

√ **c.** scissors : cut

_____ **d.** rock : hard

_____ **e.** grass : green

1. Electric generator : electricity ::

_____ **a.** heater : blanket

_____ **b.** heating : window

_____ **c.** freezing : ice

_____ **d.** freezer : ice

_____ **e.** plants : photosynthesis

2. Power plant : fossil fuels ::

_____ **a.** nuclear plant : nuclear fuels

_____ **b.** power plant : radioactive material

_____ **c.** fossil fuel : biomass

_____ **d.** resources : water

_____ **e.** electricity : wire

3. nonrenewable resource : limited ::

_____ **a.** biomass : heat

_____ **b.** nuclear energy : nuclear plant

_____ **c.** oil : coal

_____ **d.** solar energy : disadvantage

_____ **e.** renewable resource : abundant

4. Fission : splitting ::

_____ **a.** chain reaction : neutron

_____ **b.** fusion : combining

_____ **c.** nuclear plant : radioactive waste

_____ **d.** geothermal energy : drilling

_____ **e.** hydrothermal energy : dam

5. Nuclear fusion : star ::

_____ **a.** electricity : electric generator

_____ **b.** combustion : car

_____ **c.** steam : biomass

_____ **d.** sunlight : solar cells

_____ **e.** nuclear fission : heat

6. solar energy : solar cell ::

_____ **a.** power plant : fossil fuel

_____ **b.** mechanical energy : electric generator

_____ **c.** nuclear fusion : nuclear fission

_____ **d.** solar energy : solar heating

_____ **e.** liquid fuels : biomass

7. Liquid fuel : biomass ::

_____ **a.** hydrogen : liquid

_____ **b.** oil : coal

_____ **c.** ethanol : fruit

_____ **d.** gasohol : gas

_____ **e.** solar cells : sunlight

79

CRITICAL THINKING WORKSHEET

INTERPRETING OBSERVATIONS

Read the following scenario, and then answer the questions below.

Imagine that you are a civil engineer who was just hired by the government of Iceland to plan a city. On the airplane you read about the "land of fire and ice," and learn that it was formed from tectonic activity. As a result, there are many volcanoes, geysers, and hot springs on the island. Also, because Iceland is located in northern latitudes, it receives a lot of snowfall in winter.

As your plane zooms in for a landing, you notice that the annual springtime snowmelt has formed large lakes and colossal waterfalls—two of which are very impressive! Some volcanoes lining an overcast sky are still capped with snow. You notice that there aren't very many trees around, only grass and shrubs in a rocky landscape.

1. What are two sources of energy you would use for your city? Explain your answer.

_____

_____

_____

_____

_____

2. Using materials already at your disposal, how could you construct the houses so that they are well-insulated? How could you heat the water?

_____

_____

_____

_____

3. Would it be effective to use solar cells or solar collectors on the roofs to provide electricity or hot water? Explain your answer.

_____

_____

_____

_____

_____

Name _____ Class _____ Date _____

AGREE OR DISAGREE

Agree or disagree with the following statements, and support your answer.

1. Producing electricity on a large scale inevitably has environmental costs.

_____

_____

_____

_____

_____

_____

2. Electricity should cost more money so that people will use less.

_____

_____

_____

_____

_____

_____

_____

3. Fifty years from now, our major source of energy will be sunlight.

_____

_____

_____

_____

_____

_____

_____

CRITICAL THINKING WORKSHEET

INTERPRETING DATA

Examine the data in the table below, and answer the questions that follow.

	How energy is used worldwide	How energy is used in the United States
Electricity generation	19%	29%
Industry	28%	19%
Transportation	16%	25%
Commercial, public, residential	17%	15%
Other	20%	12%

1. Use the data in the table above to compare how energy is used worldwide with how it is used in the United States. What recommendations would you make to city officials in the United States to encourage energy conservation? Support your answer with data from the table above.

_____

_____

_____

_____

_____

2. The United States uses 29% of its energy for electricity generation and 19% for industry, while the percentages used worldwide are nearly opposite. Explain what might account for this difference. Hint: consider the two groups into which most countries fall.

_____

_____

_____

_____

_____

_____

_____

CRITICAL THINKING WORKSHEET

READING COMPREHENSION AND ANALYSIS I

Read the following passage, and answer the questions that follow.

Consumers are not interested in buying raw energy, but in getting access to energy services—heat, light, hot water, etc. If those services can be provided more cheaply through efficiency than by generating more energy, then it is just good business for utilities to supply those services, and to enable their consumers to use energy more efficiently.

L. Hunter Lovins*

1. How might efficiency lower an energy provider's costs?

_____

_____

_____

_____

2. Do you agree with the author's statement that "consumers are not interested in buying raw energy?" Explain your answer.

_____

_____

_____

_____

_____

3. What problems, if any, do you see with providing energy more efficiently?

_____

_____

_____

_____

_____

_____

*From "Efficiency: Less Energy, More Power" by L. Hunter Lovins in *Environment*, 2nd edition. Copyright © 1998 by **Saunders College Publishing, a division of Harcourt Brace & Company.** Reprinted by permission of the publisher.

CRITICAL THINKING WORKSHEET

READING COMPREHENSION AND ANALYSIS II

Read the following passage, and answer the questions that follow.

Tidal power utilizes the twice daily changes in sea levels to generate electricity. It is suitable only for certain coastlines, but it certainly offers great possibilities in places where the tides vary twenty to a hundred feet per day. Wave power is another dynamic area awaiting development.

Helen Caldicott*

1. What does this passage generally imply about options for providing us with energy?

_____

_____

_____

_____

_____

_____

2. How is the author using the word *dynamic* in this passage?

_____

_____

_____

_____

_____

3. As a scientist, how would you investigate the energy-supplying potential of tides?

_____

_____

_____

_____

_____

*From *If You Love This Planet* by Helen Caldicott. Copyright © 1992 by **W. W. Norton & Company, Inc.**
Reprinted by permission of the publisher.

CHAPTER REVIEW

CHAPTER
12

WASTE

Matching   Match each example in the left column with the appropriate term from the right column.

_____ **1.** household rubbish

_____ **2.** landfill water containing hazardous waste

_____ **3.** waste containing lead, mercury, or cadmium

_____ **4.** newspaper, cotton fibers, leather

_____ **5.** polyester, nylon, plastic

_____ **6.** steam, ash, electrostatic precipitator

a. biodegradable materials

b. synthetic materials

c. leachate

d. municipal solid waste

e. hazardous waste

f. incinerator

Concept Mapping

7. Demonstrate your understanding of municipal solid waste management. For each example below, draw a line from each material to the appropriate solid-waste management technique to each possible outcome.

Materials	Waste management techniques	Possible outcomes
paper bag	recycling	rich soil
		combustible gases such as methane
plastic utensils	incinerator	leachate seepage into groundwater
leaves	landfill	air pollution
		debris goes to landfill
batteries	compost	reused as another material

Multiple Choice In the space provided, write the letter of the word or statement that best answers the question or completes the sentence.

_____ 1. Solid waste includes all of the following EXCEPT

    a. agricultural waste.

    b. methane.

    c. plastics.

    d. food waste.

_____ 2. The ash produced by incinerators is _____ than other solid waste.

    a. less toxic

    b. more toxic

    c. as toxic

    d. more recyclable

_____ 3. The law that makes owners of hazardous-waste sites responsible for cleanup is

    a. the Superfund Act.

    b. the Incinerator Act.

    c. the EPA.

    d. the Love Canal Act.

_____ 4. Surface impoundment relies on which physical process?

    a. decontamination

    b. compression

    c. evaporation

    d. interjection

_____ 5. Discarded material that is not in liquid or gas form is technically called

    a. crud.

    b. junk.

    c. sludge.

    d. solid waste.

_____ 6. The amount of yard waste produced in the United States in 1995 was about

    a. 45,000 tons.

    b. 20,000 tons.

    c. 10,000 tons.

    d. 30,000 tons.

_____ 7. The best way to make sure radioactive waste does not pose a serious threat to humans is to

    a. treat it with chemicals.

    b. store it where it can decay far away from people and water sources.

    c. deep-well inject it.

    d. decontaminate it.

_____ 8. The amount of municipal solid waste going to landfills could be reduced by

    a. composting yard waste.

    b. recycling.

    c. reusing products.

    d. All of these answers are correct.

Name _____ Class _____ Date _____

Short Answer   Write the answers to the following questions in the spaces provided.

1. Define recycling, and provide two examples.

_____

_____

_____

_____

_____

_____

_____

2. Explain how plastic bags can be labeled "biodegradable" when plastic is nonbiodegradable.

_____

_____

_____

_____

_____

_____

_____

_____

_____

_____

3. Explain why some types of waste are considered hazardous waste, and provide examples.

_____

_____

_____

_____

_____

_____

_____

_____

CRITICAL THINKING WORKSHEET

ANALOGIES

Mark the letter of the pair of terms that best completes the analogy shown. An analogy is a relationship between two pairs of words or phrases written as a:b::c:d. The symbol : is read *is to,* and the symbol :: is read *as*.

Example
keyboard : type ::

_____ **a.** plane : land

_____ **b.** dog : eat

√ **c.** scissors : cut

_____ **d.** rock : hard

_____ **e.** grass : green

1. solid waste : leachate ::

_____ **a.** tree : sap

_____ **b.** pie : filling

_____ **c.** liquid waste : toxins

_____ **d.** body : sweat

_____ **e.** salty soil : salty water

2. Oil : recycling ::

_____ **a.** battery : community collection site

_____ **b.** solid waste : hazardous waste

_____ **c.** compost : leaves

_____ **d.** waste reduction : material conversion

_____ **e.** compost : nylon

3. Pond : surface impoundment ::

_____ **a.** deep-well injection : drill

_____ **b.** baby : diapered baby

_____ **c.** oil : groundwater

_____ **d.** hazardous waste : incineration

_____ **e.** degradable : biodegradable

4. Leachate : groundwater ::

_____ **a.** synthetic material : compost

_____ **b.** energy : heat

_____ **c.** plastic : landfill

_____ **d.** oil : drilling

_____ **e.** ash : air

5. Hazardous waste: deep-well injection ::

_____ **a.** solid waste : landfill

_____ **b.** solid waste : waste management

_____ **c.** leachate : landfill

_____ **d.** surface impoundment : compost

_____ **e.** biodegradable material : heat

6. radioactive : hazardous ::

_____ **a.** leachate : impoundment

_____ **b.** biodegradable : solid

_____ **c.** bottles : plastic

_____ **d.** compost : hazardous

_____ **e.** solid : waste

7. Cotton: biodegradable ::

_____ **a.** incinerator : ash

_____ **b.** yard waste : nonbiodegradable

_____ **c.** polyester : nonbiodegradable

_____ **d.** over-packaged : not packaged

_____ **e.** bomb : flammable

88

CRITICAL THINKING WORKSHEET

CHAPTER
12

THINKING ABOUT PROCESSES

Read the following scenario, and answer the questions that follow.

In an ideal world, tires sent for recycling would be ground into small particles and reformed into new tires. But it is actually not that simple. Tires are made through a process, called *vulcanization*, in which raw rubber is heated and combined with sulfur to make a sturdy, elastic compound. Without the addition of sulfur, rubber would have the consistency of clay. However, the sulfurized rubber is so chemically stable that chemical bonds cannot form between tire particles and added rubber, so old tires cannot be easily converted into new ones. Currently, the only widely-used technology for recycling tires involves processing them with dangerous chemicals, such as chlorine and sulfur dioxide.

1. Based on what you have learned in this chapter, what might be some of the disadvantages of processing tire rubber with chemicals?

   _____

   _____

   _____

   _____

2. We rely heavily on microbes to decompose organic wastes in landfills and compost heaps. There is one type of bacteria, *Sulfolobus acidocaldarius,* that thrives only on sulfur in Yellowstone's hot springs. How might this bacteria help us recycle tires?

   _____

   _____

   _____

3. Scientists use the term *biodesulfurization* to describe the process explored in question 2. Based on the words in this term and on the scenario above, what do you think biodesulfurization means?

   _____

   _____

   _____

   _____

CRITICAL THINKING WORKSHEET

CHAPTER

12

AGREE OR DISAGREE

Agree or disagree with the following statements, and support your answer.

1. Leachate is a form of hazardous waste.

_____

_____

_____

_____

_____

2. Compost is biodegradable.

_____

_____

_____

_____

_____

_____

_____

3. There is little danger of hazardous waste entering groundwater if the waste is disposed of through deep-well injection.

_____

_____

_____

_____

_____

_____

_____

Name _____ Class _____ Date _____

READING COMPREHENSION AND ANALYSIS

Read the following passage, and answer the questions that follow. Also read page 423 of the textbook for an explanation of the role of economics in environmental decision making, including the decision of whether or not to pollute.

The rational man finds that his share of the cost of the wastes he discharges into the commons is less than the cost of purifying his wastes before releasing them. Since this is true for everyone, we are locked into a system of 'fouling our own nest,' so long as we behave only as independent, rational, free-enterprisers.

Garrett Hardin*

1. What idea is the author trying convey? Hint: The *commons* means resources that we all share or hold in common, such as clean air, water, and soil.

   _____

   _____

   _____

   _____

2. What does this passage imply about the difference between independent and collective rational behavior?

   _____

   _____

   _____

3. How can we as a society prevent "fouling our nest?"

   _____

   _____

   _____

   _____

   _____

   _____

   _____

*From "The Tragedy of the Commons" by Garrett Hardin from *Science,* vol. 162, December 13, 1968, pp. 1243–1268. Copyright © 1968 by the **American Association for the Advancement of Science.** Reprinted by permission of the publisher.

CRITICAL THINKING WORKSHEET

CHAPTER 12

READING COMPREHENSION AND ANALYSIS II

Read the following passage, and answer the questions that follow.

You don't get these types of incinerators and chemical plants being compatible with clean industries or office towers. To create white-collar office jobs you have to attract the population, and usually people like to live near where they work.

When the toxic landfill came into Warren County, North Carolina, giving birth to the environmental justice movement, the county started to lose major businesses, because people started to identify the county with hazardous wastes.

Robert Bullard*

1. What happens when a community that produces hazardous wastes refuses to store them?

_____

_____

_____

_____

2. Do you agree with the author's idea that incinerators and chemical plants are not "compatible" with clean industries or office towers? Why?

_____

_____

_____

3. Imagine that a hazardous-waste producer in a neighbor state has proposed storing its waste in your community. How do you think your community would react?

_____

_____

_____

_____

_____

_____

*From "Some People Don't Have 'The Complexion for Protection'" by Robert Bullard from *E: The Environmental Magazine,* vol. IX. no. 4, July/August 1998. Copyright © 1998 by **Earth Action Network.** Reprinted by permission of the publisher.

CHAPTER REVIEW

POPULATION GROWTH

Matching Match each example in the left column with the appropriate term in the right column.

_____ **1.** maximum growth rate of a population

_____ **2.** effect of factors that limit population growth

_____ **3.** movement of organisms out of a population

_____ **4.** factors that can limit population growth

_____ **5.** people that often change locations to find food

_____ **6.** transition from hunting and gathering to cultivating plants and raising animals

a. limiting resources

b. hunter-gatherers

c. emigration

d. biotic potential

e. agricultural revolution

f. environmental resistance

Graphing

7. A population of rabbits moves to a new habitat with a high carrying capacity. Several generations later a severe drought reduces the carrying capacity of the land. Complete the population growth curve below to show how the rabbit population responds to these changes.

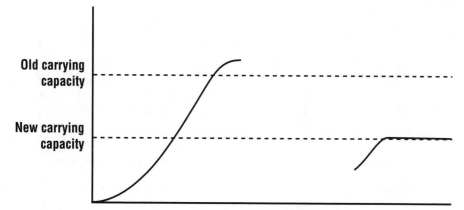

Population Growth Curve

CHAPTER 13 REVIEW, CONTINUED

Multiple Choice   In the space provided, write the letter of the word or statement that best answers the question or completes the sentence.

_____ 1. The graph of human population growth over time since 1200 A.D. looks like

    a. a J-curve.

    b. an S-curve.

    c. a horizontal line.

    d. a straight 45° line.

_____ 2. Plant populations are limited by all of the following EXCEPT

    a. water.

    b. nesting sites.

    c. mineral nutrients.

    d. light.

_____ 3. A population will shrink if deaths + emigrants exceed

    a. deaths + births.

    b. immigration − emigration.

    c. births + immigrants.

    d. the carrying capacity.

_____ 4. Developed countries tend to have relatively even numbers

    a. of limiting resources.

    b. of economic opportunities.

    c. of people in each age group.

    d. urban problems.

_____ 5. Environmental resistance might decrease with the following:

    a. a higher biotic potential

    b. a larger population

    c. more immigrants

    d. more food

_____ 6. About how long did it take the human population to double from 2 billion to 4 billion people?

    a. 130 years

    b. 45 years

    c. 95 years

    d. 175 years

_____ 7. A population's biotic potential represents its growth rate

    a. under ideal conditions.

    b. with limiting resources.

    c. at 6% per year.

    d. without deaths.

_____ 8. Overconsumption in developed nations affects

    a. energy consumption.

    b. global resources.

    c. waste production.

    d. All of these answers are correct.

CHAPTER 13 REVIEW, CONTINUED

Understanding Trends   Read the following statements, and choose the correct answer.

_____ **1.** A J-curve represents a population that is
    a. increasing.   **b.** decreasing.   **c.** remaining the same.

_____ **2.** As shown in Figure 13-7, the human population _____ around 1100 A.D.
    a. increased   **b.** decreased   **c.** remained the same

_____ **3.** Populations that exceed the carrying capacity of their environment tend to
    a. increase.   **b.** decrease.   **c.** remained the same.

_____ **4.** According to Figure 13-8, the amount of time needed by the human population
    to double _____ until about 1987.
    a. increased   **b.** decreased   **c.** remained the same

_____ **5.** According to the theory of demographic transition, pictured in Figure 13-10,
    populations in Stage 1 tend to
    a. increase.   **b.** decrease.   **c.** remain the same.

_____ **6.** During Stage 2 of a population's demographic transition, the death rate
    a. increases.   **b.** decreases.   **c.** remains the same.

_____ **7.** When you are healthy, the growth rate of bacteria in your intestine tends to
    a. increase.   **b.** decrease.   **c.** remain the same.

_____ **8.** As shown in Figure 13-5, populations that have reached the carrying
    capacity of their environment usually do not
    a. increase.   **b.** decrease.   **c.** remain the same.

_____ **9.** Figure 13-12 shows that the population in developing countries is
    a. increasing.   **b.** decreasing.   **c.** remaining the same.

_____ **10.** The population pyramid in Figure 13-13 shows that the odds of survival in
    developed countries tend to _____ over the first four decades of life.
    a. increase   **b.** decrease   **c.** remain the same

_____ **11.** A population with a large number of emigrants and a low biotic potential is probably
    a. increasing.   **b.** decreasing.   **c.** remaining the same.

_____ **12.** Biotic potential will _____ as a population increases in size.
    a. increase   **b.** decrease   **c.** remain the same

CHAPTER 13 REVIEW, CONTINUED

Short Answer Write the answers to the following questions in the spaces provided.

1. Explain the change in lifestyle that we often call the *agricultural revolution*.

_____

_____

_____

_____

_____

_____

_____

_____

2. Compare and contrast Stage 1 and Stage 3 of the demographic transition.

_____

_____

_____

_____

_____

_____

_____

_____

3. Explain what an S-curve is and what it illustrates.

_____

_____

_____

_____

_____

_____

_____

Name _____ Class _____ Date _____

AGREE OR DISAGREE

Agree or disagree with the following statements, and support your answer.

1. We could apply the term *environmental refugees* to other animal species as well as to humans.

_____

_____

_____

_____

_____

_____

_____

_____

2. Population-related problems are primarily the concern of developing countries.

_____

_____

_____

_____

_____

_____

_____

_____

3. There should be a limit on the amount of natural resources used per individual in developed countries.

_____

_____

_____

_____

_____

_____

_____

_____

CRITICAL THINKING WORKSHEET

PROBLEM SOLVING

Read the following problems, and solve them in the spaces provided.

Math hints: Calculators can help us work with exponents. Use decimals to represent portions of 100 percent. In other words, to represent a population with a 3 percent growth rate use 1.03. The 1 represents 100 percent of the current population, and the .03 represents the 3 percent growth rate. To calculate the percent that this population would grow over several years, take this growth rate to the power of the number of years in question, then multiply by 100. So to calculate the percent growth over five years, you would write $(1.03)^5 = 1.16 \times 100\% = 116\%$. If the population originally had 100 individuals, it would have 116 individuals after five years. Most calculators use a "y^x" key to calculate exponents. In this case, you would type 1.03, y^x, 5, = to get 1.16.

1. Six golden toads, three males and three females, begin a new population in a pond in the Peruvian Andes. The population doubling time in this new environment is 4 months. Assuming the population grows at its biotic potential, how long will it take the population to reach 192 organisms?

2. On January 1, 1980, a country has 200 million people and an annual population growth rate of 8 percent. Over time, the growth rate falls, averaging 6 percent over the next 10 years. The growth rate then remains at 4 percent for the next 10 years. How large is the country's population on January 1, 2000? Show your work, and round your answer to the nearest million people.

Name _____ Class _____ Date _____

READING COMPREHENSION AND ANALYSIS I

Read the following passage, and answer the questions that follow.

Like all living things, people have an inherent tendency to multiply geometrically—that is, the more people there are the more people they tend to produce. In contrast, the supply of food rises more slowly, for unlike people, it does not increase in proportion to the existing rate of food production. This is, of course, the familiar relationship described by Malthus that led him to conclude that the population will eventually out-grow the food supply (and other needed resources), leading to famine and mass death. The problem is whether other countervailing forces will intervene to limit population growth and to increase food supply.

Barry Commoner*

1. Describe the relationship that Malthus noticed between the growth of the human population and its food supply.

   _____

   _____

   _____

2. What events have helped us avoid the mass death that Malthus predicted?

   _____

   _____

   _____

   _____

3. Is it reasonable to assume that these forces will enable us to avoid future population growth problems?

   _____

   _____

   _____

   _____

C R I T I C A L   T H I N K I N G   W O R K S H E E T

Reading Comprehension and Analysis II

Read the following passage, and answer the questions that follow.

People are perceived as poor if they eat millet (grown by women) rather than commercially produced and distributed processed foods sold by global agribusiness. They are seen as poor if they live in self-built housing made from natural material like bamboo and mud rather than in cement houses. They are seen as poor if they wear handmade garments of natural fiber rather than synthetics. Subsistence, as culturally perceived poverty, does not necessarily imply a low physical quality of life.

Vandana Shiva*

1. In your own words, what is the author saying in this passage?

_____

_____

_____

_____

2. How might living at a subsistence level be the most adaptive response to the pressures people feel in many developing nations?

_____

_____

_____

_____

3. Why do you think that many people believe that living at a subsistence level is equivalent to having a low quality of life?

_____

_____

_____

_____

_____

*From "Development, Ecology, and Women" by Vandana Shiva from *Healing the Wounds,* edited by Judith Plant. Copyright © 1992 by Judith Plant. Reprinted by permission of ***New Society Publishers.***

CHAPTER REVIEW

CHAPTER
14

TOWARD A SUSTAINABLE FUTURE

Matching   Match each example in the left column with the appropriate term from the right column.

_____ **1.** public expresses views on policy

_____ **2.** an unenforceable agreement

_____ **3.** the world's first national park

_____ **4.** blueprint for protecting the environment
and promoting sustainable development

_____ **5.** conference to develop international agreements
on the environment

_____ **6.** required by the federal government

a. nonbinding

b. EIS

c. hearings

d. Earth Summit

e. Yellowstone

f. Agenda 21

Concept Mapping

7. Complete the unfinished diagram below to illustrate the connections between the different
components.

```
            National
          Environmental
         Policy Act (NEPA)
                |
                v
          [            ]
                |
            requires
             filing of
                |
          [            ]  --written-->  [            ]
                |           by              |
          _____|_____              reviewed by
         |             |          _____|_____
    which includes     |         |      |      |
    ___|___            |     [      ][      ][      ]
   |       |           |
[      ] environmental [      ]
         impact
```

CHAPTER 14 REVIEW, CONTINUED

Multiple Choice   In the space provided, write the letter of the word or statement that best answers the question or completes the sentence.

_____ 1. Which tends to be most responsive to citizen input?

   a. local government

   b. state government

   c. federal government

   d. an international agency

_____ 2. Which is NOT an international environmental agreement?

   a. Agenda 21

   b. IWC

   c. RAMSAR

   d. Law of the Sea

_____ 3. We can use scientific knowledge and methods to

   a. help make informed environmental decisions.

   b. investigate nature.

   c. influence policy.

   d. All of these answers are correct.

_____ 4. Which group influences the type of development allowed in a particular area?

   a. state agencies

   b. planning and zoning boards

   c. local governments

   d. Both b and c are correct.

_____ 5. Which agreement seeks to eliminate wetland destruction?

   a. RAMSAR

   b. CITES

   c. MARPOL

   d. Agenda 21

_____ 6. Which problems require international cooperation?

   a. overpopulation

   b. loss of biodiversity

   c. global warming

   d. all of the above

_____ 7. How many days must the public have to comment on an Environmental Impact Report?

   a. 45 days

   b. 60 days

   c. 90 days

   d. none of the above

_____ 8. A sustainable future requires that we as individuals and as a society

   a. learn about and continue to explore environmental science.

   b. carefully consider a range of values.

   c. take an active role in our communities.

   d. All of these answers are correct.

CHAPTER 14 REVIEW, CONTINUED

Short Answer   Write the answers to the following questions in the spaces provided.

1. Define and describe the purpose of an Environmental Impact Statement.

_____

_____

_____

_____

_____

_____

2. What can an individual do to affect environmental policy at the national level?

_____

_____

_____

_____

3. Explain two problems with the 1949 International Whaling Commission agreement.

_____

_____

_____

_____

_____

4. How can environmental science help us make good environmental decisions?

_____

_____

_____

_____

_____

CRITICAL THINKING WORKSHEET

ANALOGIES

Mark the letter of the pair of terms that best completes the analogy shown. An analogy is a relationship between two pairs of words or phrases written as a:b::c:d. The symbol : is read *is to,* and the symbol :: is read *as.*

Example
keyboard : type ::

_____ **a.** plane : land

_____ **b.** dog : eat

√ **c.** scissors : cut

_____ **d.** rock : hard

_____ **e.** grass : green

1. Voting : elected representative ::

_____ **a.** voting : local elections

_____ **b.** joining lobbying organizations : national policy

_____ **c.** reading EISs : planning

_____ **d.** organizing neighborhood meetings : state government

_____ **e.** driving : commuting

2. international agreements : nations ::

_____ **a.** lobbyists : national policy

_____ **b.** EIS : public

_____ **c.** game rules : baseball players

_____ **d.** public opinion : citizens

_____ **e.** legislation : policy

3. environmental science : environmental policy ::

_____ **a.** engineering : NASA

_____ **b.** thoughts : ideas

_____ **c.** education : high school

_____ **d.** medicine : health policy

4. values : decisions ::

_____ **a.** ideas : actions

_____ **b.** negotiations : agreements

_____ **c.** knowledge : ideas

_____ **d.** actions : future

_____ **e.** all of the above

5. international cooperation : environmental problems ::

_____ **a.** conferences : agreements

_____ **b.** teamwork : obstacles

_____ **c.** science : policy

_____ **d.** environment : sustainable

_____ **e.** local : national

6. Earth Day : environmental awareness ::

_____ **a.** garage sales : fliers

_____ **b.** ban : whaling

_____ **c.** EIS : environmental impact

_____ **d.** government : citizens

_____ **e.** ban : conservation

7. national policy : state policy ::

_____ **a.** state policy : national policy

_____ **b.** local policy : local politics

_____ **c.** state policy : local policy

_____ **d.** big fish : small fry

_____ **e.** all of the above

104

CRITICAL THINKING WORKSHEET

AGREE OR DISAGREE

Agree or disagree with the following statements, and support your answer.

1. Just as the state of California has created more rigorous air quality standards than are nationally mandated, an individual is obligated to do more to preserve the environment than is required by law.

_____
_____
_____
_____
_____

2. Wealthy nations should offer poor nations economic incentives to protect the environment.

_____
_____
_____
_____
_____
_____
_____

3. Because it is so difficult for local communities to coordinate their efforts, action at the local level rarely has an effect on the environment.

_____
_____
_____
_____
_____
_____
_____

CRITICAL THINKING WORKSHEET

CHAPTER
14

REFINING CONCEPTS

The statements below challenge you to refine your understanding of concepts covered in the chapter. Think carefully, and answer the questions that follow.

1. The United States and other highly developed nations have been vocal about the logging and burning of the rain forest in the Brazilian Amazon. Beyond the environmental implications—depletion of natural resources with no plan for renewal and an increase in global production of carbon dioxide—what makes this a controversial issue?

_____

_____

_____

_____

_____

_____

2. How could the widespread study of environmental science affect government policy?

_____

_____

_____

_____

_____

_____

3. How do you think the EIS system might be made more effective?

_____

_____

_____

_____

_____

_____

Name _____ Class _____ Date _____

READING COMPREHENSION AND ANALYSIS I

Read the following passage, and answer the questions that follow.

Earth Day 1970 represented a momentous breakthrough in the wall of cultural myopia that the Western World had put between itself and the Natural World. But it is also as though the rest of humanity arrived at the cosmological home of the Indigenous Peoples and, once there, failed to greet the host family properly.

Jose Barreiro*

1. What do you think the author meant by the idea of *cultural myopia?*

_____

_____

_____

_____

_____

_____

2. What ideas might represent the cosmological home of the Indigenous Peoples?

_____

_____

_____

_____

_____

3. Do you agree with the author's statement that the Western World "failed to greet the host family properly"? Why?

_____

_____

_____

_____

_____

*From "Indigenous People Are the 'Miners' Canary' of the Human Family" by Jose Barreiro from *Learning to Listen to the Land,* edited by Bill Willers. Copyright © 1991 by Island Press. Reprinted by permission of *Alexander Hoyt Associates.*

CRITICAL THINKING WORKSHEET

READING COMPREHENSION AND ANALYSIS II

Read the following passage, and answer the questions that follow.

The debate over environmental challenges cannot be reduced to assigning blame. Patterns of consumption and resource use in the industrialized countries of the North are certainly responsible for much environmental degradation in both the North and South. But rapidly growing populations, whatever their levels of consumption, also place a greater burden on resources and the environment. Ensuring sustainability will require people to make changes, in both the way they think about their environment and how they live in it.

Nafis Sadik*

1. Do you think specific groups are blamed too often for environmental problems? Why?

_____

_____

_____

_____

2. Do you agree with the author's idea that changes in thinking and lifestyle are necessary to ensure sustainability? Why?

_____

_____

_____

_____

3. How might environmental problems be handled differently if all people were required to take an environmental science course, such as this one?

_____

_____

_____

_____

_____

*From "A Population Policy for the World" by Nafis Sadik in *Environment,* 2nd edition. Copyright © 1998 by **Saunders College Publishing, a division of Harcourt Brace & Company.** Reprinted by permission of the publisher.

Answer Key Contents

ENVIRONMENTAL SCIENCE: A GLOBAL PERSPECTIVE

CHAPTER 1

Matching

Match each example in the left column with the appropriate term from the right column.

b   **1.** see a lizard basking in the sun

e   **2.** measure body temperatures of lizards before and after basking

d   **3.** graph and analyze results

c   **4.** give a talk on why lizards bask

a   **5.** state that lizards bask in the sun to warm themselves

    a. hypothesis

    b. observation

    c. communicating results

    d. organizing and interpreting data

    e. experiment

Concept Mapping

6. Based on the graph shown below, are the depicted environmental resources—coal, sunlight, and trees—renewable or nonrenewable at current rates of use?

Coal is nonrenewable; sunlight is renewable; and trees are nonrenewable.

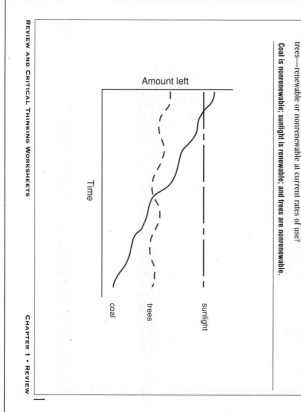

Amount left

Time

coal

trees

sunlight

Name _____ Class _____ Date _____

Graphing Connection Read the scenarios below and decide which kind of graph—line graph, pie chart, or bar graph—would best display the data. Write the name of the graph and the answers to any questions on the lines below the data tables.

1. Ted suspects that his shower has been losing pressure over the last month and wonders if his pipes are getting clogged with debris. He decides that every morning for the next week he will measure how long it takes to fill a 5 L bucket with the water going full blast. Decide which type of graph is most appropriate and whether Ted's observations support his hypothesis.

Day	Time to fill 5 L bucket (sec)
1	25
2	26
3	28
4	29
5	30
6	30
7	31

A line graph would be most appropriate. Ted's observations support his hypothesis.

2. Jill and Jerry have a vegetable farm. They are trying to decide which of two fields is better for growing zucchini and other types of squash. One field is in the shade for about half the day and the other is in full sun all day. Below are their July yields from each field. Decide which type of graph is most appropriate, and which field provides better yields.

Vegetable	Shady field	Sunny field
Zucchini	240 kg	400 kg
Acorn squash	70 kg	100 kg
Butternut squash	190 kg	290 kg

A bar graph would be most appropriate. The sunny field provides better yields.

Name _____ Class _____ Date _____

Short Answer Write the answers to the following questions in the spaces provided.

1. Explain the two different meanings of science, and provide an example of each meaning.

Science is something people know: it is the understanding about how nature works that scientists have gathered and recorded throughout human history. An example is the knowledge that the Earth revolves around the sun. Science is also something people do: it is the process of learning about nature using scientific methods. An example is an experiment to see whether a certain type of pollutant harms fish.

2. Name and describe three human activities that affect the environment.

Accept any thoughtful answer. Sample answer: 1) *Clearing land* for agricultural, residential, or industrial use destroys habitat for animals and plants and changes the ecology of the area. 2) *Burning fossil fuels*, such as gasoline and coal, releases pollutants into the atmosphere. 3) *Overconsumption* creates waste and uses nonrenewable resources.

3. Explain how gathering information contributes to making sound environmental decisions.

A person must be well-informed in order to make sound environmental decisions. Gathering information provides background about the available options, including the pros and cons and the results of any scientific studies. After gathering background information, a person may consider the values that apply to the issue, explore the possible consequences, and then make an informed decision.

CRITICAL THINKING WORKSHEET

ANALOGIES

CHAPTER 1

Mark the letter of the pair of terms that best completes the analogy shown. An analogy is a relationship between two pairs of words or phrases written as a:b::c:d. The symbol : is read *is to*, and the symbol :: is read *as*.

Example
keyboard : type ::

  a. plane : land
  b. dog : eat
✓ **c.** scissors : cut
  d. rock : hard
  e. grass : green

1. to observe : to experiment ::

  a. to repeat : to examine
  b. to test : to study
  c. to audition : to perform
✓ **d.** to watch : to manipulate
  e. real world : computer

2. pollution : poison ::

  a. industrial : natural
  b. gas : liquid
  c. cyanide : smog
✓ **d.** byproduct : intentional product
  e. harmful to environment : toxic to organisms

3. values : environmental decision ::

  a. math : economics
  b. values : scientific
  c. consequences : information
  d. weather : umbrella
✓ **e.** budget : purchase

4. biosphere : atmosphere ::

  a. organisms : weather
  b. thick : thin
  c. biology : physics
✓ **d.** concentrated : diffuse
  e. maybe : definitely

5. to sustain : to consume ::

  a. to breathe : to eat
✓ **b.** to support : to deplete
  c. to manage : to kill
  d. to generalize : to specialize
  e. to grow : to die

6. observation : hypothesis ::

  a. room is hot : like heat
  b. feel hot : like heat
  c. like heat : room is hot
✓ **d.** feel hot : have a fever
  e. have a fever : feel hot

7. ecology : environment ::

✓ **a.** organism : ecosystem
  b. interactions : surroundings
  c. animals : plants
  d. theory : science
  e. biosphere : atmosphere

CRITICAL THINKING WORKSHEET

INTERPRETING OBSERVATIONS

CHAPTER 1

Read the following scenario, and answer the questions that follow.

All summer long the weather has been very hot and dry. To conserve water, your family has not been watering the plants in your front yard. The grass is getting more yellow every day, and the bushes and flowers look pretty pathetic. However, you notice that one plant is thriving. This plant is growing in the shade of a large bush near the front door.

1. Why do you think this one plant is thriving? State your answer as a hypothesis.
Accept any reasonable hypothesis. Sample hypothesis: The plant and its underlying soil are shaded by the bush, so the plant is not drying out as rapidly as the other plants.

2. How could you test your hypothesis?
Accept any reasonable answer. Sample answer: Transplant some of the plants to a more exposed area of the yard and see how they fare. At the same time, transplant a different kind of plant from a different part of the yard into the shaded area to see if that plant then fares better.

3. What other kinds of information might you gather to help you explain this plant's success?
Accept any reasonable answer. Sample answer: Information about the plant's water requirements would be helpful. For example, one could research the identity and natural habitat of this plant (e.g., perhaps it is a desert species, more capable of surviving in dry conditions). Information about the difference between the area where the plant is growing and the rest of the yard might also explain this plant's success. One could also dig into the soil next to the thriving plant to determine whether it is getting water from a source other than rain, such as from a leaking water pipe.

CRITICAL THINKING WORKSHEET

AGREE OR DISAGREE

CHAPTER 1

Agree or disagree with the following statements, and support your answer.

1. Science is either pure or applied.

   Accept any thoughtful answer. Sample answer: Disagree; although the book distinguishes between these two forms of science, scientific knowledge and research does not always fall neatly within either category. Research that attempts to deal with a real-world problem (applied) may answer a basic (pure) question about how nature works, and vice versa.

2. Students who want to be scientists should only study science; literature and the arts have little bearing on their work.

   Accept any thoughtful answer. Sample answer: Disagree; aspiring scientists would almost surely benefit from studying literature and the arts, and their future scientific work would probably benefit as well. For example, scientists need effective writing skills to communicate their results. They could hone their writing skills by reading and writing about literature and history. For environmental scientists, other disciplines also help formulate values that may affect what they want to study and how they want to help manage our environment.

3. Most people from developing countries have values and priorities very different from those of most people from developed countries.

   Accept any thoughtful answer. Sample answer: Disagree; values and priorities will inevitably differ somewhat. However, all people, no matter what their cultural background, share the same basic needs—the need for shelter, clean water, food, and places for recreation. These needs, in turn, shape people's values. Ultimately, most people would agree that a clean, hospitable environment is a worthwhile goal for all people to work towards and share.

CRITICAL THINKING WORKSHEET

READING COMPREHENSION AND ANALYSIS 1

CHAPTER 1

Read the following passage, and answer the questions that follow.

When you can measure what you are speaking about, and express it in numbers, you know something about it; but when you cannot measure it, when you cannot express it in numbers, your knowledge is of a meager and unsatisfactory kind: it may be the beginning of knowledge, but you have scarcely, in your thoughts, advanced to the stage of *science.*

Lord Kelvin*

1. What step in the scientific method does the author refer to? Explain your answer.

   The author believes that unless you can express information quantitatively you do not have a truly scientific understanding of your subject. His point refers most specifically to organizing and interpreting data. Because most data is recorded numerically, relationships based on data can often be expressed using numbers and mathematical formulas.

2. What assumptions might the author have about the nature of knowledge and science?

   Accept any thoughtful answer. Sample answer: The author seems to assume that knowledge that is not quantifiable is somehow less useful than knowledge that is quantifiable. He also implies that scientific knowledge, specifically, is quantifiable by its very nature. Therefore, he elevates science to a level of worthiness that he presumably thinks other fields do not merit.

3. Do you agree with the author's point? Relate your answer specifically to the definition and scope of environmental science.

   Accept any thoughtful answer. Sample answer: No, I do not agree. Many observations, such as the color of a bird's wing or the appearance of a degraded landscape are not easily expressed numerically. Environmental scientists should aspire to make their work as rigorous as the work in other branches of science. However, the scope of the environmental science should extend to all aspects of our environment, many of which are not easily quantifiable.

*From a speech by William Thomson (Lord Kelvin) in 1891.

Name _____

Class _____

Date _____

CHAPTER 1

READING COMPREHENSION AND ANALYSIS II

Read the following passage, and answer the questions that follow.

In the end, we will conserve only what we love, we will love only what we understand, we will understand only what we are taught.

Baba Dioum*

1. Which do you care about more—a park near your home or an area of desert in Australia?

   Accept any answer. Chances are that students will care more about the park.

2. Does your degree of familiarity with these places influence your answer? Explain your answer.

   Accept any thoughtful answer. Sample answer: Yes. I care more about the park because of my experience there. I know the positive effect it has on my life.

3. Think of something you care about very much that your classmates may be unfamiliar with, such as a person, a pet, or a special place you like to go. Do you think your classmates would care more if they knew what you know?

   Accept any thoughtful answer. Sample answer: Yes, my classmates would care more about my pet cat, Thaddeus, if they had the good fortune to spend time with him.

4. How do the author's ideas relate to the study of environmental science?

   The author's ideas suggest that we will only learn to understand and love the environment when we are taught about it. In other words, we will only truly appreciate the natural environment when we study it, as through the field of environmental science.

*From a speech by Baba Dioum in 1968 for the International Union for Conservation of Nature and Natural Resources, New Delhi, India. Copyright © 1968 by **Baba Dioum**. Reprinted by permission of the author.

8

CHAPTER 1 • CRITICAL THINKING

HOLT ENVIRONMENTAL SCIENCE

Name _____

Class _____

Date _____

CHAPTER 2

LIVING THINGS IN ECOSYSTEMS

Matching Match each example in the left column with the appropriate term from the right column.

_e_	**1.** a single tiger shark	**a.** species
_b_	**2.** the plankton, fish, shellfish, and other organisms, plus the water, temperature, currents, and nutrients in Lake Superior	**b.** ecosystem
_c_	**3.** sunflowers in a field	**c.** population
_a_	**4.** all people	**d.** community
_d_	**5.** all of the termites, ants, beetles, mosses, and other organisms inside a rotting tree stump	**e.** organism

Concept Mapping

6. Demonstrate how the organisms shown below might interact with one another. First, draw a line from each organism to the other organism with which it might interact. Then write the name of the appropriate interaction—*predation, competition, commensalism, parasitism, or mutualism*—on the line connecting the organisms.

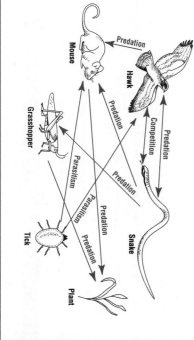

REVIEW AND CRITICAL THINKING WORKSHEETS

CHAPTER 2 • REVIEW

9

CHAPTER 2 REVIEW, CONTINUED

Multiple Choice In the space provided, write the letter of the word or statement that best answers the question or completes the sentence.

b **1.** Which is NOT a biotic factor in an ecosystem?
- **a.** weeds
- **b.** sand grains
- **c.** snakeskins
- **d.** mushrooms

d **2.** Which of the following is an example of a parasite?
- **a.** a toadstool on a tree trunk
- **b.** a cancer cell in your liver
- **c.** a bee stinger in your arm
- **d.** a flu virus in your lung

b **3.** Natural selection is not responsible for the evolution of
- **a.** the taste of mint leaves.
- **b.** the shape of a river.
- **c.** your ability to sweat.
- **d.** the shape of a fly's eye.

b **4.** Predators _____ kill their prey.
- **a.** always
- **b.** sometimes
- **c.** never
- **d.** like to

c **5.** Which of the following is an example of a species?
- **a.** German shepherds
- **b.** birds
- **c.** polar bears
- **d.** soil arthropods

d **6.** Examples of communities include all of the following EXCEPT the
- **a.** organisms in a pond.
- **b.** organisms in a fish tank.
- **c.** microbes living on you.
- **d.** dogs on your block.

a **7.** A panda bear might be all of the following EXCEPT a
- **a.** parasite.
- **b.** competitor.
- **c.** mutualist.
- **d.** predator.

b **8.** The "co-" in coevolution, as in cooperation, means
- **a.** apart.
- **b.** together.
- **c.** two.
- **d.** predator-prey.

CHAPTER 2 REVIEW, CONTINUED

Short Answer Write the answers to the following questions in the spaces provided.

1. Explain the relationship between natural selection and evolution.

According to Darwin's theory, natural selection is the unequal survival and reproduction of organisms due to the presence or absence of particular traits. Over time, natural selection changes the genetic characteristics of populations. Such changes are called evolution. Therefore, natural selection is responsible for evolution.

2. Explain the difference between a community and a population. Provide a specific example of each.

A community consists of all of the organisms living together in an ecosystem, including all of the plants and animals. An example of a community is all of the organisms living in a lake. A population, on the other hand, consists of all of the members of a single species living in a given area. An example of a population is all of the pike, a species of fish, living in a lake. Therefore, a community is simply a group of interacting populations of different species.

3. Define *competition*, and give three examples not listed in the book.

Competition is the interaction between two species that are attempting to use the same limited resource. Examples of competition include wolves and cougars preying on the same deer population, two plants drawing water from the same area of soil, and male spadefoot toads trying to attract females with which to mate.

Name _____ Class _____ Date _____

CRITICAL THINKING WORKSHEET

ANALOGIES

CHAPTER 2

Mark the letter of the pair of terms that best completes the analogy shown. An analogy is a relationship between two pairs of words or phrases written as a:b::c:d. The symbol : is read *is to*, and the symbol :: is read *as*.

Example
keyboard : type ::
___ a. plane : land
___ b. dog : cat
✓ c. scissors : cut
___ d. rock : hard
___ e. grass : green

1. Biotic factors : organisms ::
___ a. abiotic factors : plants
✓ b. abiotic factors : soil
___ c. biotic factors : physical environment
___ d. ecosystem : community
___ e. plants : water

2. Organism : habitat ::
___ a. dog : niche
___ b. giraffe : beach
___ c. bird : ecosystem
✓ d. turtle : pond
___ e. beetle : population

3. Mutualism : cooperative ::
___ a. evolution : coevolution
___ b. predator : prey
___ c. commensalism : harmful
✓ d. parenthood : nurturing
___ e. parasite : host

4. Population : community ::
✓ a. individual : population
___ b. species : organism
___ c. habitat : niche
___ d. biotic : abiotic
___ e. organism : community

5. Natural selection : evolution ::
___ a. key : keyhole
✓ b. rain : erosion
___ c. ground : landing
___ d. one : many
___ e. struggle : cooperate

6. Low : high ::
___ a. evolution : coevolution
___ b. fall : winter
✓ c. sun : earth
___ d. niche : habitat
___ e. extinct : plentiful

7. Evolution : coevolution ::
___ a. chicken : egg
✓ b. dance alone : dance with partner
___ c. ice skate : play hockey
___ d. mirror : image
___ e. playing checkers : playing chess

12

CHAPTER 2 • CRITICAL THINKING

HOLT ENVIRONMENTAL SCIENCE

Name _____ Class _____ Date _____

CRITICAL THINKING WORKSHEET

INTERPRETING OBSERVATIONS

CHAPTER 2

Read the following scenario, and answer the questions that follow.

The Venus' flytrap is a species of plant that captures flies by trapping them between hinged leaves that snap shut when a fly lands on them. Flies are attracted to the leaves because they smell like food, but it is the flies who end up being eaten!

Imagine that 1 million years ago seeds of a Venus' flytrap were blown to an island. The island had three species of flies that each lived in different areas of the island and each ate different foods. In the year 2000, scientists exploring the island discover two new species of fly-eating plants as well as the Venus' flytrap.

1. What has happened?

Two or more populations of Venus' flytraps have evolved into two new species of flytraps through the process of natural selection. These new species are each adapted to eat different species of flies.

2. Scientists call this type of evolution *adaptive radiation*. Based on the words in this term and on the scenario above, what do you think this term means?

Adaptive radiation is the evolution of different but closely-related species from a common ancestral species. The new species evolve by adapting to different circumstances from those confronted by the ancestor species, hence the term "adaptive." And the new species can be viewed as radiating out from the ancestral species, hence the term "radiation."

3. Suppose that the scientists only found one new species of flytrap. What might you conclude?

One or more populations of Venus' flytraps evolved into a new flytrap species. This could happen if two of the fly species are attracted by the same smell. It could also happen if only one population of Venus' flytraps became isolated from the other populations. If this isolated population primarily coexisted with a different fly species than did the other populations, plants that best attracted these flies would thrive and leave more off-spring. Eventually, the isolated flytrap population could have evolved into a completely new species.

REVIEW AND CRITICAL THINKING WORKSHEETS

CHAPTER 2 • CRITICAL THINKING

13

Left worksheet (page 14)

CHAPTER 2

AGREE OR DISAGREE

Agree or disagree with the following statements, and support your answer.

1. An ecosystem can be viewed as a host that is parasitized by the organisms that live in it.

   Disagree; the organisms that live in an ecosystem are an integral part of that ecosystem, while parasites are not integral parts of their hosts. Also, the organisms in an ecosystem constitute the biotic portion of the ecosystem and do not usually harm the ecosystem as a whole. However, parasites live on or in an already living host, which, by definition, they do harm.

2. The ability to unconsciously regulate your body temperature is an adaptation.

   Agree; an adaptation is an inherited trait that increases an organism's chances of survival and reproduction in a given environment. The ability to unconsciously regulate body temperature is passed from parents to offspring, so it is inherited. This ability increases the chances of surviving through summers and winters, when outside temperatures differ greatly from body temperature. Since this ability makes a person more likely to survive, it also makes a person more likely to reproduce.

3. An organism can engage in only one type of interaction (i.e. predation, mutualism, etc.) with another organism at one time. Provide examples to support your viewpoint.

   Disagree; parasitism, mutualism, and commensalism *are* mutually exclusive. If two organisms are engaged in one of these relationships, they are not, by definition, engaged in one of the others at the same time. However, organisms can engage in predator-prey interactions and competition simultaneously. For example, imported fire ants can prey on native ant species while they compete with the native species for territory. Organisms can also engage in parasitism and competition simultaneously, as when a parasitic tapeworm competes with its host for nutrients.

Right worksheet (page 15)

CHAPTER 2

REFINING CONCEPTS

The statements below challenge you to refine your understanding of concepts covered in the chapter. Think carefully, and answer the questions that follow.

1. Although there are many predators on the African savanna, none plays exactly the same role as the lion. Can any two species occupy exactly the same niche? Why or why not?

   No; within any ecosystem, each species plays a distinct role by which it gathers the resources necessary for life. Since resources, such as sunlight, food, and water, are limited, organisms that attempt to use those resources in exactly the same way will compete. Competition will lead species to evolve different strategies for acquiring resources, and, therefore, different niches within their environments.

2. Biologist Bob thinks that over time a parasite can influence the evolution of its host species. Do you think he is right? Justify your answer.

   Yes; parasites rob energy and nutrients from their hosts. For example, ticks suck blood, and tapeworms eat food that the host is digesting. Hosts with adaptations that reduce the impact of their parasites will have a better chance of surviving and reproducing (natural selection.) These adaptations will be passed to the host's offspring and eventually change the genetic characteristics of the host's population. Evolution is a change in the genetic characteristics of a population, so the parasite would have influenced the evolution of its host species through the process of natural selection.

3. One characteristic of a population is that organisms must have a reasonable chance of mating with each other. Are two wild roses separated by a wide road part of the same population? Defend your answer.

   Yes; roses mate with the help of bees and other flying insects that transfer pollen from plant to plant as they search for nectar. Since flying insects can easily cross a wide road, a rose on one side of the road has a reasonable chance of mating with one on the other side and is therefore part of the same population.

READING COMPREHENSION AND ANALYSIS I

CHAPTER 2

Read the following passage and answer the questions that follow.

Diana lives next to a large area of undeveloped oak forest. While playing in the woods with friends, she notices three species of lizards: a large, ground-dwelling species that lives on tree trunks, and a small species that lives on tree branches. All three species eat whatever small animals they can capture, usually insects.

The oak trees have recently been damaged by a type of wasp whose larvae feed on oak leaves. To save the trees, the county decides to spray the forest from the air with an insecticide. The insecticide usually does not harm lizards, but lizards can die if they eat too many poisoned wasp larvae.

1. Which of the lizards are least likely to be harmed by the insecticide? Why?

   The trunk-dwelling lizards; the poisoned wasp larvae will either be in the tree branches or on the ground (if they have died or become sick), so the lizards that live in these places are more likely to eat the larvae and become poisoned.

2. Are the lizards competing for the wasp larvae? Explain your answer.

   No; the larvae naturally live on oak leaves, so the branch-dwelling lizards are their natural predators. The trunk-dwelling and ground-dwelling lizards are adapted to prey on small organisms that spend part or all of their natural life cycles in the lizards' habitat.

3. The county sprays the trees on a still summer day. A week after the spraying, Diana learns that ducks living in a pond just up the mountain have become sick. Diana thinks she knows why. What might Diana be thinking? Is her reasoning reasonable?

   Diana is probably thinking that the ducks became sick by drinking water or eating insects tainted with the insecticide. However, her reasoning is not very reasonable. The insecticide would first have to reach the ducks, presumably by one of two ways: by wind or by water. However, the county sprayed on a calm day so it is unlikely that the wind blew the insecticide to the ducks. And since the ducks live upstream, the water could not have carried insecticide or poisoned insects to them. Therefore, it is very unlikely that the insecticide sickened the ducks.

READING COMPREHENSION AND ANALYSIS II

CHAPTER 2

Read the following passage, and answer the questions that follow.

The wolf is tied by subtle threads to the woods he moves through. His fur carries seeds that will fall off, effectively dispersed, along the trail some miles from where they first caught his fur. And miles distant is a raven perched on the ribs of a caribou the wolf helped kill ten days ago, pecking like a chicken at the decaying scrap of meat. A smart snowshoe hare that eluded the wolf and left him exhausted when he was a pup has been dead for a year now, food for an owl. The den in which he was born one April evening was home to porcupines last winter.

Barry Holstun Lopez*

1. What ideas from Chapter 2 does the author deal with in this passage?

   The author deals with the idea of the niche by describing the relationships the wolf has with its environment. The author also covers predation and competition by mentioning the wolf's interactions with a caribou, a snowshoe hare, and an owl.

2. Explain what the author means by the wolf being "tied by subtle threads to the woods he moves through."

   This statement refers to the interconnectedness of the members of an ecosystem. The wolf's niche affects all parts of its ecosystem.

3. Scientists have long battled over the idea of niches. Do niches exist? What are they? How should we think of them? Given what you have learned, explain how the niche concept is useful and how it is not useful in helping you understand an organism's way of life.

   Accept any thoughtful answer. Sample answer: the niche concept is useful in that it emphasizes the connections between all of an organism's actions. However, the niche concept is abstract because it is difficult to simultaneously picture all of these connections. The concept may be most useful as a framework for organizing observations about how an organism lives.

*From Of Wolves and Men by Barry Holstun Lopez. Copyright © 1978 by Barry Holstun Lopez. Reprinted by permission of Sterling Lord Literistic, Inc.

Name _____ Class _____ Date _____

CHAPTER REVIEW

HOW ECOSYSTEMS WORK

CHAPTER 3

Matching

Match each example in the left column with the appropriate term from the right column.

___ **1.** herbivore

___ **2.** carnivore

___ **3.** producer

___ **4.** omnivore

___ **5.** decomposer

a. oak tree

b. raccoon

c. spider

d. mushroom

e. aphid

Concept Mapping

6. Draw an energy pyramid using the following organisms found in the temperate forest of North America: shrubs, trees, bears, insects, woodpecker, hawk, rabbit, cougar, and deer. Beside your diagram, explain why you placed these animals as you did.

hawk,
shark
These tertiary consumers are carnivores and receive the least amount of energy.

woodpecker, bear
These secondary consumers are carnivores and therefore eat both herbivores and producers.

grasshopper, deer, rabbit
These primary consumers are omnivores and therefore feed on producers.

grass, shrub, tree
These producers get their energy directly from the sun.

Name _____ Class _____ Date _____

CHAPTER 3 REVIEW, CONTINUED

Short Answer

Write the answers to the following questions in the spaces provided.

1. Explain how cellular respiration fits into the carbon cycle.

Organisms break down compounds that contain carbon (organic compounds) during cellular respiration. The carbon combines with the oxygen in air to form carbon dioxide, which organisms release as a byproduct of cellular respiration. The carbon in carbon dioxide is then available for uptake by plants, which incorporate the carbon into organic compounds through photosynthesis.

2. Describe the role of decomposers in the nitrogen cycle.

Decomposers break down dead organisms and other organic matter. During this process, the decomposers release nitrogen into the soil where it can then be processed by bacteria into forms that can be used by plants or released into the atmosphere.

3. Describe the sequence of events that takes place after an agricultural field is abandoned. Do both primary and secondary succession occur? Why or why not?

When a field is abandoned, pioneer species, such as grasses and weeds, move in. Taller plants, such as perennials, then replace the pioneer plants. Eventually these plants are overshadowed by still taller plants, such as tall shrubs and small trees, which deprive the previous inhabitants of light. Finally, large trees take over, blocking out the sun for smaller trees and shrubs. Only secondary succession occurs on an abandoned field because an (agricultural) ecosystem previously existed on the soil.

CRITICAL THINKING WORKSHEET

ANALOGIES

CHAPTER 3

Name _____ Class _____ Date _____

Mark the letter of the pair of terms that best completes the analogy shown. An analogy is a relationship between two pairs of words or phrases written as a:b::c:d. The symbol : is read *is to*, and the symbol :: is read *as*.

Example

keyboard : type ::

   a. plane : land

   b. dog : cat

✓ **c.** scissors : cut

   d. rock : hard

   e. grass : green

1. producer : consumer ::

   a. car : driver

✓ **b.** factory : shopper

   c. deer : wolf

   d. photosynthesis : decomposition

   e. plants : water

2. herbivores : omnivores ::

   a. photosynthesis : respiration

   b. elephant : ocean

   c. fruit : bird

✓ **d.** deer : bear

   e. bacteria : population

3. plant : terrestrial ecosystem ::

   a. evolution : time

   b. predator : prey

   c. algae : respiration

✓ **d.** sea urchin : marine ecosystem

   e. parasite : host

4. lake : water cycle ::

✓ **a.** coal deposit : carbon cycle

   b. species : organism

   c. rain : clouds

   d. biotic : ecosystem

   e. organism : community

5. oxygen : cellular respiration ::

   a. cup : saucer

✓ **b.** carbon dioxide : photosynthesis

   c. plants : adaptation

   d. needle : thread

   e. disk : computer

6. climax forest : clear-cut forest ::

   a. plants : animals

   b. food web : food chain

   c. sun : fire

   d. ecosystem : habitat

✓ **e.** full : empty

7. trophic level : energy ::

   a. car : fuel

   b. food chain : food web

✓ **c.** grade : demerits

   d. energy pyramid : organisms

   e. taxes : income

20   CHAPTER 3 • CRITICAL THINKING   HOLT ENVIRONMENTAL SCIENCE

CRITICAL THINKING WORKSHEET

REFINING CONCEPTS

CHAPTER 3

Name _____ Class _____ Date _____

The statements below challenge you to refine your understanding of concepts covered in the chapter. Think carefully, and answer the questions that follow.

1. Do you think the distinction between the concepts of primary and secondary succession is useful? Explain your answer.

Accept any reasonable answer. Sample answer: Primary succession is the regular pattern of ecological changes that occur over time on a previously unoccupied surface. Secondary succession, on the other hand, describes the ecological changes that occur on a previously occupied surface. The distinction between the two is probably not very useful because seeds and organisms can blow and fly onto previously unoccupied surfaces (technically "primary"), and previously occupied surfaces may be biologically barren after events such as a very intense fire (technically "secondary").

2. If the sun were to burn out, explain what you think would happen to the water, carbon, and nitrogen cycles.

The water cycle would slow and eventually stop. Water would stop moving through its cycle because the there would be no solar energy to cause evaporation, to power photosynthesis, and to move air masses. The carbon cycle would also stop because it is driven by photosynthesis, which could not take place without sunlight. The nitrogen cycle would probably stop as well. Although some nitrogen could continue to move between the soil and the atmosphere without the help of plants (i.e. through processing by bacteria), the ground would eventually freeze without the warming effect of the sun, and the bacteria would presumably freeze as well.

REVIEW AND CRITICAL THINKING WORKSHEETS   CHAPTER 3 • CRITICAL THINKING   21

Name _____ Class _____ Date _____

CHAPTER 3

AGREE OR DISAGREE

Agree or disagree with the following statements, and support your answer.

1. There would be no life on Earth without the sun.

Accept any reasonable answer. Sample answers: Agree: the sun is the ultimate source of energy for all living things, even those that eat only meat. Disagree: life would continue in the deep sea, fueled by the bacteria that convert hydrogen sulfide to food. Eventually, new life-forms that build off of this new "ultimate" energy source may evolve.

2. Our activities do not upset the carbon cycle.

Disagree: by burning fossil fuels, we release carbon into the atmosphere in the form of carbon dioxide. Due to the large amounts of fossil fuels that we are burning, there are increased concentrations of carbon dioxide in the atmosphere. This upsets the previous balance between the carbon underground and in organisms and the carbon that exists in the air as carbon dioxide.

3. Urban development negatively affects the water cycle.

Agree: urban development reduces the number of plants and amount of soil available to "capture" precipitation. Therefore, more precipitation runs directly into bodies of water or evaporates from streets and sidewalks. The increased runoff causes problems with flooding that would otherwise not occur. It also reduces the amount of water that enters aquifers through recharge zones, thereby reducing the amount of groundwater available for human use.

Name _____ Class _____ Date _____

CHAPTER 3

READING COMPREHENSION AND ANALYSIS I

Read the following passage, and answer the questions that follow.

This thumbnail sketch of land as an energy circuit conveys three basic ideas:

1) that land is not merely soil;
2) that the native plants and animals kept the energy circuit open; others may or may not;
3) that man-made changes are of a different order than evolutionary changes, and have effects more comprehensive than is intended or foreseen.

These ideas, collectively, raise two basic issues:
Can the land adjust itself to the new order?
Can the desired alterations be accomplished with less violence?

Aldo Leopold*

1. Explain how the author's statements apply to the flow of energy in the environment.

The author is saying that land in its natural state keeps the flow of energy moving through ecosystems and that changing the land can alter this flow.

2. Given your new understanding of how ecosystems work, do you think land-based ecosystems as we know them will "adjust . . . to the new order"? Explain your answer.

No, it is probably not realistic. Changes to land that affect the flow of energy and materials disrupt cycles that we only partially understand. We do know that most species cannot evolve fast enough to adjust to the rapid pace of environmental change. Changes that disrupt ecological communities, such as the clearing of large areas of land, will probably keep communities at early successional states. Think of the "weed patches" that cover vacant lots. It is unlikely that ecosystems will adjust their workings to suit our purposes.

*From A Sand County Almanac by Aldo Leopold. Copyright © 1949, 1977 by **Oxford University Press, Inc.** Reprinted by permission of the publisher.

CHAPTER 3

READING COMPREHENSION AND ANALYSIS II

Name _____ Class _____ Date _____

Read the following passage, and answer the questions that follow.

The soil exists in a state of constant change, taking part in cycles that have no beginning and no end. New materials are constantly being contributed as rocks disintegrate, as organic matter decays, and as nitrogen and other gases are brought down in rain from the skies. At the same time other materials are being taken away, borrowed for temporary use by living creatures. Subtle and vastly important chemical changes are constantly in progress, converting elements derived from air and water into forms suitable for use by plants. In all these changes living organisms are active agents.

Rachel Carson*

1. How can the soil exist "in a state of constant change" when it continues to play the same role in the cycling of materials?

The soil does play the same role, as with the fixation of nitrogen, but it plays this role through the active participation of living organisms, through the gradual breakdown of rocks, and the breakdown of dead organisms and organic waste matter. The soil is able to maintain its role only through its continuous activity; it is very much dynamic, not static.

2. What does the author mean by "cycles that have no beginning and no end"? Explain your answer.

The author means by cycles that continue indefinitely. The movement of water, carbon, and nitrogen through the environment will not stop as long as there is sunlight to power the cycles.

*From Silent Spring by Rachel Carson. Copyright © 1962 by Rachel L. Carson; copyright renewed © 1990 by Roger Christie. All rights reserved. Reprinted by permission of Houghton Mifflin Company.

24 CHAPTER 3 • CRITICAL THINKING

CHAPTER 4

KINDS OF ECOSYSTEMS

Name _____ Class _____ Date _____

Matching Match each example in the left column with the appropriate term from the right column.

e	1. desert	a. cold and dry
a	2. tundra	b. fertile soils
d	3. forest	c. Mediterranean climate
b	4. grassland	d. canopy vegetation
c	5. chaparral	e. hot and dry

Concept Mapping

6. Complete the climatograms below based on what you have learned about the climate of these different biomes. Vancouver is located in a temperate rain forest, which is characterized by cool, moist winters. The city receives about 130 cm of precipitation per year, a summer high of 18°C, and a winter low of 2°C. Buffalo's climate is typical for a northern deciduous forest. Use a line for temperature and fill in the bar graph for precipitation. It is not important for your answers to be exact but for you to demonstrate that you understand how these climatograms would generally look.

TEMPERATE RAIN FOREST
(Vancouver, British Columbia)

PRECIPITATION (CM) / TEMPERATURE (°C)
MONTHS: J F M A M J J A S O N D

TEMPERATE DECIDUOUS FOREST
(Buffalo, New York)

PRECIPITATION (CM) / TEMPERATURE (°C)
MONTHS: J F M A M J J A S O N D

CHAPTER 4 • REVIEW 25

121

CHAPTER 4 REVIEW, CONTINUED

Multiple Choice   In the space provided, write the letter of the word or statement that best answers the question or completes the sentence.

a

1. Thin soil, high temperatures, and high rainfall represent a
 a. tropical rain forest.
 b. temperate rain forest.
 c. desert.
 d. grassland.

b

2. Birds migrating in winter, coniferous plants, and cold temperatures represent a
 a. South Pole.
 b. taiga.
 c. temperate forest.
 d. chaparral.

c

3. Eutrophication, littoral zone, and zooplankton represent a
 a. grassland.
 b. coral reef.
 c. lake.
 d. photosynthesis.

c

4. Harbor, phytoplankton, and high productivity represent a
 a. marsh.
 b. river.
 c. estuary.
 d. benthic zone.

c

5. The North and South Poles can both be considered
 a. terrestrial (land-based) ecosystems.
 b. part of the taiga.
 c. marine ecosystems.
 d. home to penguins.

b

6. All of the following are examples of freshwater ecosystems EXCEPT
 a. swamps.
 b. estuaries.
 c. marshes.
 d. the Everglades.

a

7. Productive ecosystems include
 a. estuaries and rain forests.
 b. tundra and savanna.
 c. taiga and desert.
 d. coral reefs and lakes.

b

8. Factors that influence which plants grow where include
 a. longitude.
 b. climate.
 c. biome maps.
 d. None of these answers are correct.

CHAPTER 4 REVIEW, CONTINUED

Short Answer   Write the answers to the following questions in the spaces provided.

1. Explain how the soil of the tropical rain forest can support the most plant species of any biome yet contain so few nutrients.

The nutrients in rain forests are contained primarily within living plants. Dead organisms decay quickly because of the heat and moisture, and the plants absorb most of the nutrients from these decaying organisms. Much of the few remaining nutrients are washed away by the constant rains.

2. Describe the role of fire in chaparral and grassland biomes.

During the dry season, fires periodically burn through chaparral and grassland communities. The fire kills many trees, leaving the fire-adapted grasses and chaparral plants to sprout and take over. Fire allows the characteristic chaparral and grassland plant species to outcompete plants that would otherwise eventually grow large enough to deprive native species of sunlight. Fire also returns nutrients to the soil, providing nourishment for the plants that sprout.

3. Describe the difference between the benthic and the littoral zones, and include the organisms you would expect to find in each.

The benthic zone is the bottom of a body of water. Decomposers, insect larvae, and clams live in the benthic zone. The littoral zone is the nutrient-rich area along the shore of lakes and ponds. Along the littoral zone you would expect to find insects, rooted plants, and small fish.

CRITICAL THINKING WORKSHEET

ANALOGIES

CHAPTER 4

Mark the letter of the pair of terms that best completes the analogy shown. An analogy is a relationship between two pairs of words or phrases written as a:b::c:d. The symbol : is read *is to*, and the symbol :: is read *as*.

Example
keyboard : type ::
  a. plane : land
  b. dog : eat
  ✓ **c.** scissors : cut
  d. rock : hard
  e. grass : green

1. Shallow : deep ::
  a. tundra : taiga
  b. littoral : benthic
  c. ecosystem : biome
  d. teacher : student
  e. marsh : swamp

2. Hibernation : cold ::
  a. ocean : estuary
  b. estivation : hot
  ✓ **c.** wet : freezing
  d. dry : desert
  e. fur : winter

3. Bromeliads : trees ::
  a. plants : soil
  b. fish : water
  c. fleas : dogs
  d. water : land
  ✓ **e.** sponges : coral reef

4. Trees : canopy ::
  ✓ **a.** bushes : understory
  b. weeds : bushes
  c. saplings : trees
  d. birds : flock
  e. roots : plants

5. Polar bear : tundra ::
  a. cactus : desert
  ✓ **b.** moose : taiga
  c. desert : cactus
  d. bison : savanna
  e. lobster : wetland

6. Phytoplankton : zooplankton ::
  a. algae : lichens
  b. reeds : fish
  c. little fish : big fish
  d. coral : shrimp
  ✓ **e.** plants : animals

7. Salty : brackish ::
  a. swamp : river
  ✓ **b.** hot : warm
  c. thermal : heat
  d. saline : salt
  e. cool : cold

CRITICAL THINKING WORKSHEET

INTERPRETING DATA

CHAPTER 4

Examine the following data, and answer the questions that follow.
Plant #1: broad leaves, leaves turn yellow in autumn, tall
Plant #2: waxy coating, spines, long and shallow root system
Plant #3: needlelike leaves, pyramid shape, likes acidic soil

1. What is plant #1, which biome is it from, and how is it adapted to that biome?

Plant 1 is a deciduous tree found in a temperate forest. Its broad leaves help it absorb energy in the summer while its height shades out competition. The leaves turn colors and drop in autumn so that the tree can better survive the long, cold winters of temperate forests.

2. What is plant #2, which biome is it from, and how is it adapted to that biome?

Plant 2 is a cactus. Its succulent flesh stores water, allowing it to weather the hot, dry desert climate. Its photosynthetic stems have a waxy cover to preserve water and spines to protect it from predators. The cactus's long, shallow root system helps it collect as much water as possible during the desert's infrequent rains.

3. What is plant #3, which biome is it from, and how is it adapted to that biome?

Plant 3 is a coniferous tree found in the taiga. Its needlelike leaves help it to preserve water during the winter, and its pyramidal shape helps it shed heavy snow. Another adaptation is the conifer's ability to affect the soil through its needles. When its needles drop, they acidify the soil surrounding the conifer, making the soil inhospitable for plants that might otherwise grow near the conifer and compete with it for water, nutrients, and sunlight.

CRITICAL THINKING WORKSHEET

AGREE OR DISAGREE

CHAPTER 4

Agree or disagree with the following statements, and support your answer.

1. Many types of ecosystems can exist within a given region.

   Agree; biome describes the general plant and animal community that dominates a given biome. Different types of deciduous forest are all considered temperate deciduous forest. Also, biomes contain ecosystems that are not included in the overall biome profile. For example, deserts contain arroyos (wetlands), and temperate forests and taiga contain fields (grasslands).

2. Humankind's conversion of grasslands to croplands was necessary.

   Agree; while many of the native grasslands of our planet have been destroyed with the introduction of farmland, this change was crucial to the success of humans as a species. To support the agriculture of our growing populations, we needed to use the most fertile land. The United States was particularly successful at grassland agriculture, but it has cost us the majority of the grassland itself.

3. Coral reefs are the marine equivalent of tropical rain forests.

   Agree; coral reefs share many similarities with tropical rain forests. Both biomes have very high productivity, meaning they produce a lot of biomass. Both biomes are also characterized by a high level of biodiversity. Finally, both of these biomes face major threats from development and pollution now and in the foreseeable future.

CRITICAL THINKING WORKSHEET

REFINING CONCEPTS

CHAPTER 4

The statements below challenge you to refine your understanding of concepts covered in the chapter. Think carefully, and answer the questions that follow.

1. Recommend a strategy for incorporating sustainable human activity into a biome.

   Accept any reasonable answer. Sample answer: An example of a sustainable strategy for an estuary is to allow limited fishing, swimming, and recreational boating but to limit shoreline development in order to protect water quality. Money from tourism could support the local economy and also be used to purchase more land to preserve and maintain areas already set aside.

2. Is artificial eutrophication undesirable considering that eutrophication occurs naturally anyway? Defend your answer.

   Artificial eutrophication is undesirable because it changes the balance of nutrients, and, therefore, the ecology of the system as a whole. Although eutrophication will eventually occur naturally, excess nutrients enable algae and other organisms to grow unchecked and effectively smother the lake. Artificial eutrophication creates a crash-and-burn situation that can kill most organisms in the lake—a situation from which the area may recover only very slowly.

3. What would happen if a conifer from the taiga was planted in a tropical rain forest? (Consider roots, growth, and the above ground portions in your answer.)

   The conifer would have difficulty sending roots as deep as it normally does because the ground underneath the thin tropical soil is hard. Also, the roots would find few nutrients at this deeper level, so the tree would begin to starve. At the same time, the conifer might drown in the wet tropical climate because its needles are adapted to conserve water in the taiga, where it rains less frequently and the dry air evaporates moisture from leaves.

Left Worksheet

Name _____ Class _____ Date _____

READING COMPREHENSION AND ANALYSIS I

CHAPTER 4

Read the following passage and answer the questions that follow.

As in deserts, a limiting physical factor rules these lands (the tundra), but it is heat rather than water that is in short supply in terms of biological functioning. Precipitation is also low, but water as such is not limiting because of the low evaporation rate. Thus, we might think of the tundra as an arctic desert, but it can best be described as a wet arctic grassland or a cold marsh that is frozen for a portion of the year.

Eugene Odum*

1. In your own words, what is the author saying?
The author is saying that the tundra can be best understood as a biome adapted to very cold conditions in the same way that the desert is best understood as a biome adapted to very dry conditions.

2. What does the author mean by limiting factors?
Limiting factors are essential requirements, such as sufficient sunlight, water, or nutrients, whose limited supply restricts the type of community that can live in a given area.

3. Does the author blur the distinction between different biomes (tundra, grasslands, marshes), or does he simply use them to draw an analogy? Defend your answer.
Accept any reasonable response. Sample answer: The author does not blur the distinction between different biomes because different biomes still share common characteristics. For example, tropical rain forests and temperate deciduous forests are both dominated by tall trees, just as the tundra and marshes are both wet most of the year. The author is simply drawing an analogy to help the reader understand the tundra's ecology.

*From *Ecology: A Bridge Between Science and Society* by Eugene P. Odum. Copyright © 1997 by **Sinauer Associates, Inc.** Reprinted by permission of the publisher.

32
CHAPTER 4 • CRITICAL THINKING
HOLT ENVIRONMENTAL SCIENCE

Right Worksheet

Name _____ Class _____ Date _____

READING COMPREHENSION AND ANALYSIS II

CHAPTER 4

Read the following passage, and answer the questions that follow.

The house knows the sound of El Río Grande; river that for centuries wandered through this Chihuahua desert, largest desert in North America, old ocean bed where millions of years ago, land emerged from water, mountains rise. Oceans became seas, seas dried to lakes, and lakes evaporated into basins and playas. Water creatures—oysters, clams, coral—hardened in the sea of sand, wordless geological history.

Pat Mora*

1. What point is the author making about the permanence of biomes?
The author is pointing out that biomes change over long periods of time. In any given area, the biome that exists today probably did not exist several million years ago.

2. Should this point have an effect on our efforts to preserve today's biomes? Justify your response.
Accept any thoughtful response. Sample answer: no, this point should not affect today's preservation efforts. Biomes change naturally over very long periods of time, while humans live a relatively short period of time. In other words, the changes we make today, good or bad, will probably affect ourselves, our children, and our grandchildren.

3. What effect do you think geological history—the changes of the Earth's surface over time—has on the rise and fall of different biomes?
The movement and activity of the Earth's surface have a great deal to do with the rise and fall of biomes. Changes in the Earth's surface affect the location and elevation of land areas, which in turn affect the climate of those areas. Climate largely determines what can live where, which determines the biome present in any given area at any given time.

*From *House of Houses* by Pat Mora. Copyright © 1997 by Pat Mora. Reprinted by permission of **Beacon Press.**

33
REVIEW AND CRITICAL THINKING WORKSHEETS
CHAPTER 4 • CRITICAL THINKING

Name _____ Class _____ Date _____

WATER

CHAPTER 5

Matching Match each example in the left column with the appropriate term from the right column.

e   **1.** desalinization

b   **2.** aquifer

d   **3.** land-based

a   **4.** bacteria

c   **5.** multiple sources

f   **6.** native plants

a. pathogen
b. recharge zone
c. nonpoint pollution
d. ocean pollution
e. reverse osmosis
f. water conservation

True/False   Decide whether the following statements are true or false, and place a T or F in the space to the left of each statement.

F   **1.** Approximately 0.2 percent of all water on Earth is groundwater.

F   **2.** Heat energy can be a form of water pollution.

T   **3.** Nonpoint pollution is a special form of point pollution.

F   **4.** Polluting in a recharge zone could contaminate an aquifer.

T   **5.** Biological magnification and artificial eutrophication can both result from water pollution.

F   **6.** Dams affect ecosystems both upstream and downstream.

F   **7.** Most of the oil polluting the oceans comes from major oil spills.

T   **8.** Estuaries bear the brunt of the effects of ocean pollution.

T   **9.** The rate of groundwater recharge affects the time it takes to decontaminate aquifers.

10. In wastewater treatment plants, sedimentation tanks provide the time it takes to decontaminate aquifers. the necessary environment for the aerobic decomposition of sewage.

Name _____ Class _____ Date _____

Short Answer   Write the answers to the following questions in the spaces provided.

1. Explain the relationship between an aquifer and its recharge zone.

An aquifer is a large underground rock formation that holds groundwater. Water enters the ground as precipitation in areas called recharge zones. Therefore, recharge zones provide the groundwater for aquifers.

2. Define thermal pollution, and provide an example.

Thermal pollution is the addition of heat energy into bodies of water at levels that are unhealthy for the animals and plants that live there. An example of thermal pollution is the dumping of warm water from an industrial plant into a river.

3. Compare and contrast point and nonpoint pollution.

Point pollution is pollution that originates from a single source, such as a factory. Nonpoint pollution, on the other hand, is pollution that comes from many different sources. An example of nonpoint pollution is the oil that washes into bodies of water from driveways and streets.

4. Explain why only a small portion of the Earth's water is available for human use.

The vast majority of the Earth's water is unavailable for human use for two main reasons. First, 97 percent of the Earth's water is salt water, which we cannot use to drink or to irrigate crops. Second, more than two thirds of the Earth's fresh water is frozen in icecaps and glaciers.

Name _____ Class _____ Date _____

CRITICAL THINKING WORKSHEET

ANALOGIES

CHAPTER 5

Mark the letter of the pair of terms that best completes the analogy shown. An analogy is a relationship between two pairs of words or phrases written as a:b::c:d. The symbol : is read *is to*, and the symbol :: is read *as*.

Example
Keyboard : type ::
___ a. plane : land
___ b. dog : eat
✓ c. scissors : cut
___ d. rock : hard
___ e. grass : green

1. Dam : reservoir ::
✓ a. wreck : traffic jam
___ b. braces : teeth
___ c. river : dam
___ d. groundwater : aquifer
___ e. heat : heat wave

2. Artificial eutrophication : fertilizer ::
___ a. pollution : accumulation
___ b. point pollution : nonpoint pollution
___ c. biological magnification : pollution
✓ d. evolution : selection
___ e. salinization : salt

3. Water : watershed ::
___ a. food : stomach
✓ b. groundwater : aquifer
___ c. blood : arm
___ d. river : lake
___ e. iron : blood

4. Pathogen : *Escherichia coli* ::
___ a. organism : biotic
___ b. pathogen : feces
✓ c. parasite : flea
___ d. fish : organism
___ e. bacteria : disease

5. Nonpoint pollution : multiple sources ::
___ a. pesticides : salt
✓ b. point pollution : single source
___ c. biotic factors : pathogen
___ d. aquifer : recharge zone
___ e. eutrophication : succession

6. Desalination : salt ::
___ a. pumping : water
___ b. pumping : aquifer
✓ c. pollution : decontamination
___ d. dehydration : water
___ e. eutrophication : sewage

7. Water pollutant : pathogen ::
___ a. thermal pollution : physical
✓ b. vegetable : broccoli
___ c. point pollution : street oil
___ d. septic tank : point pollution
___ e. pathogens : DDT

Name _____ Class _____ Date _____

CRITICAL THINKING WORKSHEET

AGREE OR DISAGREE

CHAPTER 5

Agree or disagree with the following statements, and support your answer.

1. A river is greater than the sum of the streams that feed it.

Agree; a river drains the entire land area that constitutes its watershed. When it rains, much of the water never enters streams but flows over the ground's surface and even underground toward a river. Aquifers also contribute to rivers in some areas.

2. Ocean pollution is more likely to affect a large sea animal, like a whale, than a small shoreline animal, like a crab.

Disagree; since most ocean pollution originates on land and enters the ocean at or near the shore, shoreline animals are more likely to be exposed to concentrated pollutants than large sea animals, which usually live in the open ocean.

3. Pumping and diverting water affects the water cycle in predictable ways.

Accept any thoughtful answer. Sample answer: Disagree; although we have a fairly good understanding of how surface and groundwater systems work, we cannot predict how the weather will change from week to week. Since much of the water cycle occurs in the atmosphere, and we cannot predict how our actions affect weather and climate, we cannot know with any precision how pumping and diverting water will affect the water cycle.

Worksheet 1 (left)

Name _____ Class _____ Date _____

THINKING SCIENTIFICALLY

CHAPTER 5

Read the following scenario, and answer the questions that follow.

On your way to school you notice that the water in a previously clear local stream now appears brown. You have heard that a new paper-processing plant recently opened upstream, and you wonder if there might be a connection.

1. Form a hypothesis based on your observations.

Accept any reasonable explanation that can be tested. Sample hypothesis: The brown color of the water below the new plant is caused by the plant's activities.

2. How would you test your hypothesis and evaluate your results?

Accept any reasonable response. Sample answer: Observe and take samples of the water at several points upstream and downstream from the new plant. Create an index for comparing differences in water color. For example, a score of "0" could represent totally clear water, and a score of "5" could represent black water, with intermediate scores representing various shades of brown. Score your samples based on the index. Plot your data on a bar graph with locations on the x-axis and the scores on the y-axis, and look for trends.

3. Assume that your data seem to contradict your hypothesis. What can you conclude?

One can conclude that the new plant is not the source of the brown coloring that you observed.

4. How might you construct a new hypothesis based on the information you collected?

Accept any reasonable response. Sample answer: One could pinpoint the section of river that seems to be the source of the discoloration and form a hypothesis based on whatever activities are going on along that area of stream.

Worksheet 2 (right)

Name _____ Class _____ Date _____

READING COMPREHENSION AND ANALYSIS

CHAPTER 5

Read the following passage, and answer the questions that follow.

Everything depends on the manipulation of water—on capturing it behind dams, storing it, and rerouting it in concrete rivers over distances of hundreds of miles. Were it not for a century and a half of messianic effort toward that end, the West as we know it would not exist.

Marc Reisner*

1. The word *messianic* is the adjective form of the word *messiah.* A messiah is an expected savior or liberator of a people or country. Why do you think the author used this word to describe the effort to manipulate water in the West?

The author chose this word because the manipulation of water in the West was intended to "save" the West from its lack of fresh water.

2. In your own words, what is the author saying?

The author is saying that the urban and suburban southwestern United States would not exist without various human-made water works. And the lushness of places like Palm Springs and Phoenix would be impossible without the importation and pumping of enormous quantities of water.

3. What do you think is the "everything" that the author refers to in the first line?

The "everything" that the author refers to is the physical state of our society today—comforts such as running water at any time, green lawns without rain, fresh fruit year-round, and so on. He is saying that the shape of the modern human world is one that does not conform to the physical limits set by nature.

4. Does your understanding of water supply issues in the western United States support the author's claim? Explain.

Accept any reasonable answer. Sample answer: Yes, as explained in the book, several western states draw so heavily on the Colorado River that it is almost exhausted by the time it reaches Mexico. Also, most of the southwestern United States is composed of desert and chaparral, biomes that receive relatively little rainfall. It seems unlikely that the natural climate of these biomes could support the civilization that now exists there.

*From *Cadillac Desert,* revised and updated, by Marc P. Reisner. Copyright © 1986, 1993 by Marc P. Reisner. Reprinted by permission of **Viking Penguin, a division of Penguin Putnam Inc.***

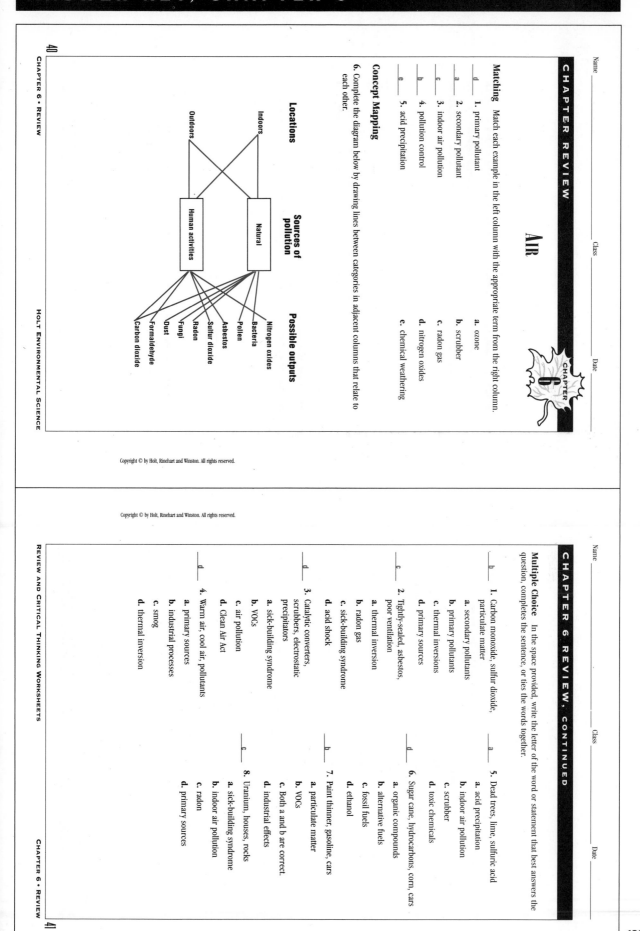

CHAPTER 6

AIR

Matching Match each example in the left column with the appropriate term from the right column.

___d___ **1.** primary pollutant

___a___ **2.** secondary pollutant

___c___ **3.** indoor air pollution

___b___ **4.** pollution control

___e___ **5.** acid precipitation

a. ozone

b. scrubber

c. radon gas

d. nitrogen oxides

e. chemical weathering

Concept Mapping

6. Complete the diagram below by drawing lines between categories in adjacent columns that relate to each other.

Locations

Indoors

Outdoors

Sources of pollution

Human activities

Natural

Possible outputs

Nitrogen oxides

Bacteria

Pollen

Asbestos

Sulfur dioxide

Radon

Fungi

Dust

Formaldehyde

Carbon dioxide

Multiple Choice In the space provided, write the letter of the word or statement that best answers the question, completes the sentence, or ties the words together.

___b___ **1.** Carbon monoxide, sulfur dioxide, particulate matter

a. secondary pollutants

b. primary pollutants

c. thermal inversions

d. primary sources

___c___ **2.** Tightly-sealed, asbestos, poor ventilation

a. thermal inversion

b. radon gas

c. sick-building syndrome

d. acid shock

___d___ **3.** Catalytic converters, scrubbers, electrostatic precipitators

a. sick-building syndrome

b. VOCs

c. air pollution

d. Clean Air Act

___d___ **4.** Warm air, cool air, pollutants

a. primary sources

b. industrial processes

c. smog

d. thermal inversion

___a___ **5.** Dead trees, lime, sulfuric acid

a. acid precipitation

b. indoor air pollution

c. scrubber

d. toxic chemicals

___d___ **6.** Sugar cane, hydrocarbons, corn, cars

a. organic compounds

b. alternative fuels

c. fossil fuels

d. ethanol

___b___ **7.** Paint thinner, gasoline, cars

a. particulate matter

b. VOCs

c. Both a and b are correct.

d. industrial effects

___c___ **8.** Uranium, houses, rocks

a. sick-building syndrome

b. indoor air pollution

c. radon

d. primary sources

Name _____ Class _____ Date _____

CHAPTER 6 REVIEW, CONTINUED

Short Answer   Write the answers to the following questions in the spaces provided.

1. Name and describe three effects of air pollution on health.
   Accept any reasonable answer. Sample answer: One effect of air pollution on health is chronic bronchitis, a persistent inflammation of the bronchial linings. A second effect is asthma, a condition in which the bronchial passages constrict and become blocked with mucus. A third effect is emphysema, the loss of elasticity in the lung's air sacs.

2. How is mass transit in cities a possible solution to urban air pollution?
   Though many cities are compact, people still need to travel fairly long distances to get to work and school, to run errands, and so on. Mass transit, including the use of buses and subways, reduces the amount of fuel burned per passenger. This reduces the amount of pollutants that are released into the air. Therefore, the more people who use mass transit versus private vehicles, the less pollutants enter the air and the cleaner our urban air will be.

3. Describe two atmospheric conditions that make air pollution worse.
   One condition that makes air pollution worse is a thermal inversion, which occurs when warm air high in the atmosphere traps cooler, polluted air beneath it. A second condition that worsens air pollution is the trapping of polluted air in valleys. The polluted air does not blow away because it is trapped between hills or mountains.

Name _____ Class _____ Date _____

INTERPRETING OBSERVATIONS

Read the following scenario, and answer the questions that follow.

Lake Sulfox seems to be having some problems with its fish population. Commercial fisherman are claiming that their catches have declined, and they are blaming the decline on the supposed acidification of the lake by a local coal-fired power plant. The Lake Sulfox Advisory Board has the following data on file. Assume that the size of the fish harvest is a good indicator of the size of the fish population.

Annual Fish Harvest (metric tons)

1991	1992	1993	1994	1995	1996
7500	6924	6322	5412	5503	5113

Mean Sulfate Levels (ppm)

1991	1992	1993	1994	1995	1996
41.07	51.34	54.89	57.46	58.76	59.65

1. What is the relationship between the size of the fish harvest and the sulfate levels in the lake?
   The higher the sulfate levels in the lake, the lower the size of the fish harvest.

2. Do the data prove that acidification of the lake by sulfates is responsible for the decline in the lake's fish population? Provide at least two reasons to support your viewpoint.
   No, the tests do not prove that acidification is causing the decline of the fish population for two reasons. One reason is that there is no proof that sulfate levels affect fish populations. A second reason is that there is no way to know that the fish harvest would not have declined without the addition of sulfates to the lake. In other words, without more data it is impossible to know if the fish harvest was already declining, presumably for other reasons.

Name _____ Class _____ Date _____

CHAPTER 6

REFINING CONCEPTS

The statements below challenge you to refine your understanding of concepts covered in the chapter. Think carefully, and answer the questions that follow.

1. Imagine that you are the city manager, and the EPA has given your city a citation for high levels of oxides in the air. What steps would you take to reduce the levels of oxides in the air?

One solution is to ensure that mass transit is available and convenient for many people. This would reduce the use of individual automobiles. Another solution could be to offer incentives to carpool, such as lower high-way tolls and designated carpool lanes. Finally, increase enforcement of the laws that govern air pollution, such as requiring yearly emissions testing.

2. Where might using grain as an alternative fuel source not be a good idea? Explain.

Using grain as an alternative fuel source might not be a good idea for developing countries. Many of these countries desperately need their grain supplies to feed their human populations.

3. Manufacturing is often blamed for producing air pollution. However, according to economic theory, it is neither technologically feasible nor economically efficient to completely eliminate pollution. What do you think? Explain your reasoning.

Accept any reasonable response. Sample answers: Agree; it would be impossible to maintain our society without any pollution. Many businesses would fail, and many people would lose their jobs if businesses were banned from producing any pollution. Even non-industrial societies produce pollution in the form of smoke from fires. It makes more sense for our society to work toward drastically reducing pollution rather than completely eliminating it.

44 CHAPTER 6 • CRITICAL THINKING HOLT ENVIRONMENTAL SCIENCE

Name _____ Class _____ Date _____

CHAPTER 6

READING COMPREHENSION AND ANALYSIS

Read the following passage, and answer the questions that follow.

If the wind can carry seeds and spores and the fragrance of hay, it can also carry man-made molecules, the gases and compounds exhaled by industrial activities. We have, in fact, depended on it to do so, building higher smokestacks in hopes that the air currents farther up will carry our smoke farther away, out of sight, out of mind, eliminating them as a local problem, at least.

Stephanie Mills[*]

1. Do you agree with the author's idea that air pollution is often an "out of sight, out of mind" problem? Explain your answer.

Accept any thoughtful answer. Sample answer. Yes; people seem reluctant to solve air pollution problems until the problems affect the quality of air in their community. Stories about air pollution in cities such as Mexico City have done little to stimulate public action on a large scale.

2. How do you think limiting the height of smokestacks would affect the air pollution problem?

Accept any thoughtful answer. Sample answer. If air pollution was not carried away by air currents, then the industries would be polluting local communities. Eventually, people might be motivated to restrict industrial production of pollutants.

3. Many environmentalists have proposed selling a limited number of permits to facilities that release pollutants into the air. In theory, any person or group could buy the permits. How could this help solve the air pollution problem?

Accept any thoughtful answer. Sample answer. With a limited number of permits, the total amount of pollution would be restricted. Private citizens or groups would be able to purchase clean air by purchasing but not using permits.

[*]From *In Praise of Nature* by Stephanie Mills. Copyright © 1990 by Island Press. Reprinted by permission of *Alexander Hoyt Associates.*

REVIEW AND CRITICAL THINKING WORKSHEETS CHAPTER 6 • CRITICAL THINKING 45

Name _____ Class _____ Date _____

ATMOSPHERE AND CLIMATE

CHAPTER 7

Matching Match each example in the left column with the appropriate term from the right column.

e 1. increased snow cover
c 2. CFCs in atmosphere
h 3. ocean warming
b 4. burning fossil fuels
d 5. low-angle sunlight
a 6. high levels of ozone in upper atmosphere
f 7. increased carbon dioxide in atmosphere
g 8. high UV radiation at Earth's surface

a. less UV radiation reaches Earth's surface
b. more CO_2 in atmosphere
c. ozone destruction
d. cooler temperatures
e. more sunlight reflected from Earth
f. heat is trapped near surface
g. increased DNA damage
h. more water vapor in atmosphere

Concept Mapping

9. The climate at the equator is wet and rainy but very dry at latitudes 30 degrees north and south of the equator. Climate is the average weather pattern of a region over time. Complete the concept map below to illustrate the processes responsible for a region's climate. Use the following terms: *air circulation, precipitation, ocean currents, latitude, local geography, Earth's rotation, solar energy received, and winds.*

Climate **is determined by**
latitude / air circulation / local geography / ocean currents

latitude **which affects** → solar energy received, precipitation

ocean currents **which are caused by** → winds, Earth's rotation

46 CHAPTER 7 • REVIEW HOLT ENVIRONMENTAL SCIENCE

Name _____ Class _____ Date _____

Multiple Choice In the space provided, write the letter of the statement that best answers the question or completes the sentence.

a 1. The weather we experience occurs in the
a. troposphere.
b. stratosphere.
c. ozone layer.
d. exosphere.

d 2. Rain is common whenever
a. cold, moist air rises.
b. warm, moist air rises.
c. warm, dry air sinks.
d. cold, dry air sinks.

c 3. Earth's atmosphere is the only place in the solar system with large amounts of
a. water vapor.
b. methane.
c. oxygen.
d. weather.

d 4. Cloud cover makes the ground-level temperature
a. cooler by day.
b. cooler by night.
c. warmer by night.
d. Both a and c are correct.

d 5. The gas most responsible for the greenhouse effect is
a. nitrous oxide.
b. methane.
c. oxygen.
d. water vapor.

b 6. Which of the following reduce CO_2 in the atmosphere?
a. phytoplankton
b. tropical rain forests
c. oceans
d. All of the answers are correct.

c 7. In the summer, the northern hemisphere gets sunlight
a. obliquely for long days.
b. slanting for short days.
c. more directly for long days.
d. less directly for short days.

c 8. Ozone in the stratosphere
a. causes skin cancer.
b. prevents DNA repair.
c. absorbs UV light.
d. destroys CFCs.

REVIEW AND CRITICAL THINKING WORKSHEETS CHAPTER 7 • REVIEW 47

Name _____ Class _____ Date _____

Short Answer   Write the answers to the following questions in the spaces provided.

1. Human activities currently release 7 billion tons of carbon dioxide into the atmosphere each year. What is the source of this excess carbon dioxide?

   Most of the carbon dioxide released into the atmosphere comes from burning oil, gas, and coal in cars, factories, airplanes, and buildings. The rest comes from burning biomass, mainly forests.

2. What are the most important steps industrialized countries could take to help reduce levels of atmospheric carbon dioxide?

   One of the most important steps is to reduce reliance on fossil fuels for transportation. Another step is to build and encourage people to use mass transit systems, such as subways. A third step we could take is to further invest in alternative energy sources such as solar panels, hydroelectricity, and wind power for generating heat and electricity.

3. Why are some developed countries willing to pay to help developing countries find substitutes for CFCs?

   If developing countries continue to use CFCs, it will reduce the ozone layer everywhere. When developed countries stop using CFCs, the ozone layer will still be depleted if developing countries are using CFCs.

4. Why is it important for politicians to understand scientific information on the greenhouse effect and ozone holes?

   Politicians get together to make and then approve agreements to cut CFC usage or carbon dioxide emissions. If they don't have good scientific information, they might vote in a way that would damage the environment and threaten life for all of us.

Name _____ Class _____ Date _____

INTERPRETING EVIDENCE

CHAPTER 7

Read the following passage, and answer the questions that follow.

Alaska is thawing, as is much of Canada and northern Russia. Hundreds of glaciers are retreating. The warmer atmosphere has produced more snow to feed the glacier, but longer, warmer summers melt them faster than the heavier snows can build them. The region's permafrost is thawing in the interior, and pockets of underground ice trapped in the frost are melting too. Forests are drowning as the ground sinks and water floods their roots. Trees, weakened by climate-related stresses, are killed by spruce bark beetles whose population has exploded. Average global temperatures have increased by 1°F over the last century. But in Alaska, Siberia, and northwest Canada, average temperatures have increased as much as 5°F over the last 30 years. Warming is more pronounced in winter. Mainstream scientists predict that Alaska is expected to warm twice as much as the global average.

Anne Gregory*

1. After warming starts, is a snow-ice region likely to show a sharper temperature rise than a region without snow-ice? Explain your answer.

   Snow and ice melt when the temperature rises above freezing, exposing the ground. The ground absorbs much more sunlight than ice and therefore warms up more rapidly. More snow-ice then melts and more ground is exposed, further increasing the temperature. So areas with snow-ice will likely experience a sharper temperature rise than areas without snow-ice.

2. Do you agree with the mainstream scientists' prediction about Alaska? Justify your response.

   Accept any reasonable answer. Sample answer: Yes; the data so far indicates that the Arctic is already warming faster than the rest of the Earth. Because of the positive feedback loop, Alaska and other polar regions will continue to warm more rapidly than temperate and tropical regions.

*Anne Gregory, 1998

CRITICAL THINKING WORKSHEET

AGREE OR DISAGREE

CHAPTER 7

Agree or disagree with the following statements, and support your answer.

1. Industrial countries should assist tropical rain forest countries so those countries can afford to leave their forests intact.

   Accept any thoughtful answer. Sample answer: Agree; rain forest countries cut down the forest to sell the timber or to clear land to raise cattle. The forests provide a source of income for these countries. But the forests also help reduce carbon dioxide levels for the planet, and industrial countries produce most of the extra carbon dioxide. Therefore, industrial nations should help tropical countries develop economic uses for rain forests, such as tourism and medicinal sources.

2. The correlation between carbon dioxide levels in the atmosphere and world temperatures for the past 160,000 years proves that higher carbon dioxide levels cause global warming.

   Accept any thoughtful answer. Sample answer: Disagree; the correlation does not really show cause and effect. It only shows that high carbon dioxide levels occur at the same time as warmer temperatures. Perhaps a third factor, such as increased solar energy, caused both.

3. Developing countries should not participate in treaties that set allowable levels of greenhouse emissions in developed countries.

   Accept any thoughtful answer. Sample answer: Disagree; underdeveloped countries are not producing much of the greenhouse gases, but they will suffer as much as other countries from the effect of the gases that industrial countries produce and use.

HOLT ENVIRONMENTAL SCIENCE

CRITICAL THINKING WORKSHEET

REFINING CONCEPTS

CHAPTER 7

The statements below challenge you to refine your understanding of concepts covered in the chapter. Think carefully, and answer the questions that follow.

1. Some scientists predict that global warming will cause major ocean currents to shut down. The Gulf Stream moves warm water toward northern latitudes, and the South Polar Conveyor Belt moves cold water toward the Equator. How might an ocean current shutdown affect the climate?

   Ocean currents move cold and warm water around the globe. If they stop, then warm water will stay near the equator, and cold water will stay near the poles. The climate will become colder at high latitudes and warmer near the equator, ultimately evening out temperatures at all latitudes.

2. A catalyst speeds up a process but is not changed itself. CFCs are known to release catalysts that break down the ozone layer. How does this process work?

   UV radiation in the stratosphere breaks up CFC molecules, releasing chlorine atoms. The chlorine acts as a catalyst by reacting with and breaking apart ozone molecules. The chlorine atoms are not destroyed when they react with ozone, but instead continue to react with and destroy thousands of other ozone molecules in the stratosphere.

3. The carbon in fossil fuels was in the atmosphere long ago. Why is it now a problem that we burn those fossil fuels and put the carbon back into the atmosphere?

   Sample answer: Fossil fuels are the remains of plants that lived millions of years ago. The fossil fuels represent millions of years worth of photosynthetic activity that pulled enormous quantities of carbon dioxide out of the air and converted it into hydrocarbons. Now we are burning the fossil fuels and returning these millions of years of carbon storage to the atmosphere in just a few hundred years.

Name _____ Class _____ Date _____

CHAPTER 7

READING COMPREHENSION AND ANALYSIS

Read the following passage, and answer the questions that follow.

So, when the earth's atmosphere acquired its oxygen from the photosynthetic activity of green plants, the planet also acquired a high-altitude blanket of ozone. Until then the earth's surface was bathed in intense ultraviolet radiation, which was, in fact, the energy source that converted the early earth's blanket of methane, water, and ammonia into the soup of organic compounds in which the first living things originated. However, ultraviolet radiation is very damaging to the delicate balance of chemical reactions in living cells, and it is likely that the first living things survived only by growing under a layer of water sufficiently thick to protect them from the ultraviolet radiation that reached the earth's surface.

Barry Commoner*

1. How might the decrease in UV radiation reaching the Earth's surface have affected the evolution of life-forms?

   UV radiation reaching the Earth's surface burns tissue and genetically damages living cells. A decrease in UV probably allowed more-sensitive organisms to survive. Organisms were therefore able to evolve into a much greater variety of forms.

2. Read the last sentence of the passage again. How does the ozone layer differ from the water that shielded early organisms from UV light?

   The ozone layer and water both absorb UV radiation. However, stratospheric ozone is a gas that does not benefit organisms other than by absorbing UV rays. Water, on the other hand, is a liquid medium that provided the opportunities and limits within which early organisms could live and evolve.

3. Do you think the human species could survive without the ozone layer? Explain your answer.

   Our species might survive, but probably not for many generations unless we radically altered our lifestyles. Even if we moved into building complexes and underground, we would still need to eat. Our food is primarily produced on land and outdoors. Most crops would not survive the intense UV radiation that the ozone layer now filters.

*From *The Closing Circle: Nature, Man, and Technology* by Barry Commoner. Copyright © 1971 by Barry Commoner. Reprinted by permission of *Alfred A. Knopf, Inc.*

52 CHAPTER 7 • CRITICAL THINKING

HOLT ENVIRONMENTAL SCIENCE

Name _____ Class _____ Date _____

CHAPTER 8

LAND

Matching Match each example in the left column with the appropriate term from the right column.

d	1. infrastructure	a. multiple use
f	2. urbanization	b. protected land
g	3. deforestation	c. reclamation
e	4. mineral resource	d. bridges
b	5. wilderness	e. sulfur
c	6. mining	f. development
a	7. public land	g. clear-cutting

Ranking Human activity changes the land. Rank the following types of land use according to how much the landscape has been altered by humans. Rank the least changed category 1 and the most changed 7.

Land Use Type	Landscape Change Rank (1–7)
8. corn farm	5
9. national park	1
10. rangeland	3
11. suburban shopping mall	6
12. abandoned wheat field	4
13. downtown business district	7
14. managed forest	2

53 CHAPTER 8 • REVIEW

REVIEW AND CRITICAL THINKING WORKSHEETS

CHAPTER 8 REVIEW, CONTINUED

Multiple Choice   In the space provided, write the letter of the word or statement that best answers the question or completes the sentence.

___a___ **1.** Reclamation reduces the ultimate impact of
- **a.** mining.
- **b.** crime.
- **c.** overgrazing.
- **d.** air pollution.

___c___ **2.** All of the following are found in wilderness except
- **a.** habitat.
- **b.** camping.
- **c.** suburbs.
- **d.** fishing.

___b___ **3.** A modern consequence of our growing urban areas is
- **a.** lower crime rate.
- **b.** suburban sprawl.
- **c.** infrastructure.
- **d.** energy conservation.

___b___ **4.** Open pit mining creates
- **a.** ore.
- **b.** holes.
- **c.** strips.
- **d.** mineral resources.

___c___ **5.** The "wedges and corridors" plan is an example of
- **a.** architecture.
- **b.** urban crisis.
- **c.** highway construction.
- **d.** land-use planning.

___a___ **6.** Tree-harvesting methods include
- **a.** selective cutting.
- **b.** reforestation.
- **c.** clear-cutting.
- **d.** Both a and c are correct.

___a___ **7.** A fire station is an example of
- **a.** infrastructure.
- **b.** suburbanization.
- **c.** land-use planning.
- **d.** renovation.

___d___ **8.** Rangeland may suffer from
- **a.** overgrazing.
- **b.** desertification.
- **c.** poor management.
- **d.** All of the above answers are correct.

CHAPTER 8 REVIEW, CONTINUED

Short Answer   Write the answers to the following questions in the spaces below.

1. Explain the advantages and disadvantages of clear-cutting as a method of harvesting timber.

Clear-cutting is the most economical way to harvest timber in the short run because all of the trees in an area are cut for use. However, in the long run, the areas best suited for growing trees will be unavailable for cutting until a new crop of trees has grown to maturity. While clear-cutting is more cost-effective in the short run, it is often less cost-effective in the long run and is not a sustainable method of resource management.

2. Explain the role of inadequate infrastructure in the urban crisis.

Infrastructure includes all of the facilities built for public use in our cities. Recently, as more people have moved to cities all over the world, the infrastructure in place has not been adequate to provide all the services necessary. Urban problems such as traffic jams and overloaded sewer lines stem from the lack of adequate infrastructure. These problems contribute to the crisis conditions that now plague many urban areas.

3. Describe two methods for managing rangeland.

One method, grazing management, concentrates on limiting the time cows are allowed to graze the land in order to sustain the land's productivity. In this method, cows are moved frequently to avoid disturbing any one area too much. A second method, range improvement, focuses on managing the land itself. Range improvement techniques include eliminating weedy plants, planting vegetation on bare spots, fencing overgrazed areas to let them recover, and creating enough water holes that cows do not overgraze the land surrounding any one hole.

Name _____ Class _____ Date _____

Understanding Vocabulary For each pair of terms explain the difference in their meanings.

1. *Suburban sprawl* and *urban crisis*

The urban crisis is the lack of adequate infrastructure, including roads, sewers, and bridges, in modern cities. Suburban sprawl is a consequence of the urban crisis. Suburban sprawl refers to development, such as houses and strip malls, that spreads out around cities.

2. *Clear-cutting* and *selective cutting*

Both clear-cutting and selective cutting are methods of harvesting timber. Clear-cutting involves removing all the trees from a land area. Selective cutting involves cutting only mature or middle-aged trees and replacing removed trees with saplings.

3. *Deforestation* and *desertification*

Deforestation is the wholesale removal of all trees from an area. Desertification is the land degradation caused by overuse and overgrazing. Desertified land is no longer productive for growing crops, grazing farm animals, or nourishing native plants.

4. *Reforestation* and *reclamation*

Reforestation is the process of replacing trees that have died or been removed from a forest. Reclamation is the restoration of a previously mined area of land to the way it looked before it was mined.

56

CHAPTER 8 • REVIEW

HOLT ENVIRONMENTAL SCIENCE

Name _____ Class _____ Date _____

AGREE OR DISAGREE

CHAPTER 8

Agree or disagree with the following statements, and support your answer.

1. It is more economically advantageous in the long run to protect or preserve open space.

Accept any thoughtful answer. Sample answer: Agree; preserving tracts of land does decrease the amount of land that can be developed. However, people like to live and work in areas that include or are near preserved land. Therefore, developers can attract people to shop or buy homes by using land in environmentally-sustainable ways. Open space also reduces drainage problems by absorbing and slowly releasing large quantities of precipitation.

2. Mining companies should restore land to the same successional stage it was in when they mined it. Figure 3-16, on page 67, illustrates different successional stages of a forest.

Accept any reasonable answer. Sample answer: Agree; land that has been mined has suffered serious damage and may not easily recover to its former state without a lot of "help." Mining operations clear large land areas, which would be infiltrated by surrounding organisms only very slowly if it is left to recover from an early successional state.

3. The lack of adequate subway systems in most urban areas of the United States reflects poor land-use planning.

Accept any thoughtful answer. Sample answer: Agree; most urban areas were developed around the use of the car. Land-use planning was not done with the future in mind, a future that would eventually involve traffic problems without an adequate mass-transit system.

57

REVIEW AND CRITICAL THINKING WORKSHEETS

CHAPTER 8 • CRITICAL THINKING

137

Name _____ Class _____ Date _____

REFINING CONCEPTS

CHAPTER 8

The statements below challenge you to refine your understanding of concepts covered in the chapter. Think carefully, and answer the questions that follow.

1. Mining dramatically alters the form and function of land. What ecological challenges do you think a mining reclamation team might face?

Accept any reasonable answer: Sample answer: One challenge is restoring soil, including the rich topsoil that nourishes plants and usually took thousands of years to form. Another challenge is reestablishing natural water drainage, which again relates to the structure of the soil and bedrock. A reclamation team would also have to reintroduce the right combination of organisms to create the same community that was present before.

2. If you were a land-use planner, what guidelines would you use to locate a new mall on the outskirts of your community?

Accept any reasonable answer: Sample guidelines: The mall should be located near a highway so that new long roads would be unnecessary. The mall should also be located on land that is already disturbed, like an abandoned farm. Finally, the developers must survey the area to make sure that valuable natural resources, such as wetlands, are not destroyed unnecessarily.

3. What issues in your community or in the nearest city could be considered part of the urban crisis? Discuss these issues.

Accept any reasonable answer. Possible answers include constructing new highways, widening existing roads, developing new shopping centers, houses, or schools, and building a new airport. Traffic increases as a result of construction, but eventually there are new schools for kids, new businesses, and easier ways to get around the city.

58

Name _____ Class _____ Date _____

READING COMPREHENSION AND ANALYSIS I

CHAPTER 8

Read the following passage, and answer the questions that follow.

The bare vastness of the Hopi landscape emphasizes the visual impact of every plant, every rock, every arroyo. Nothing is overlooked or taken for granted. Each ant, each lizard, each lark is imbued with great value simply because the creature is there, simply because the creature is alive in a place where any life at all is precious. . . . So little lies between you and the earth. One look and you know that simply to survive is a great triumph, that every possible resource is needed, every possible ally—even the most humble insect or reptile. You realize you will be speaking with all of them if you intend to last out the year.

Leslie Marmon Silko*

1. What do you think is the author's overall message in this passage?

Accept any thoughtful response. Sample answer: The author is saying that the starkness of the Hopi landscape is such that you feel a sense of kinship, or shared fate, with the other organisms that are trying to eke out an existence there.

2. What sort of attitude does the author have toward the land and its resources?

Accept any thoughtful response. Sample answer: The author has a deep respect for the land and its many resources conscientiously. She thinks that in order to make a home there, she has to treat the land and what it offers.

3. How do we "speak" to the many land-based resources that we use?

Accept any thoughtful response. Sample answer. By using the word "speak," the author implies that we have a conversation of sorts with the land's resources. That is, we engage in a certain give and take with them. For example, we harvest timber and fish and in turn must either wait for new trees and fish to grow or help replace them. Many of our "conversations" are one-sided, such as when we take continually from an area without sowing the seeds we will need for the future.

*From "Landscape, History, and the Pueblo Imagination" by Leslie Marmon Silko from *The Nature Reader*. Copyright © 1986 by Leslie Marmon Silko. First published in *Antaeus*. Reprinted by permission of *The Ecco Press*.

59

138

Name _____ Class _____ Date _____

CRITICAL THINKING WORKSHEET

CHAPTER **8**

READING COMPREHENSION AND ANALYSIS II

Read the following passage, and answer the questions that follow.

There is yet no ethic dealing with man's relation to land and to the animals and plants which grow upon it. . . . land is still property. The land-relation is still strictly economic, entailing privileges but not obligations.

Aldo Leopold*

1. What are the "privileges and obligations" that the author is referring to?

   Accept any thoughtful answer. Sample answer: Generally, the author is saying that society has viewed land as strictly something to be used and taken advantage of but not to protect or preserve. One privilege is the land's fertility, which yields food, timber, and fiber and nourishes domestic and game animals. Another privilege is the provision of clean water. The author feels that society does not obligate people to maintain the fertility, beauty, or ecological health of the land. In other words, we do not feel obligated to have a sustainable relationship with the land.

2. Explain how we can use methods outlined in the chapter to fulfill our "obligations."

   Accept any reasonable answer. Sample answer: One method that will help us fulfill our obligations is land-use planning, which minimizes the impact of development by locating development where it has the fewest negative effects on the land. Another useful method is selective cutting, which allows us to maintain a forest population of varying ages of trees and retain the many services that intact forests provide.

3. Who do you think is responsible for instituting a land ethic? Explain your answer.

   Accept any thoughtful answer. Everyone is responsible for instituting a land ethic. Governments are responsible because they have authority. Individuals and businesses are responsible because they own land. Parents are responsible because they raise children. And schools and universities are responsible because they disseminate knowledge.

*From *A Sand County Almanac* by Aldo Leopold. Copyright © 1949, 1977 by **Oxford University Press, Inc.** Reprinted by permission of the publisher.

60 CHAPTER 8 • CRITICAL THINKING   HOLT ENVIRONMENTAL SCIENCE

Name _____ Class _____ Date _____

CHAPTER REVIEW

CHAPTER **9**

FOOD

Matching Match each example in the left column with the appropriate term from the right column.

g	1. clearing forests	**a.** less erosion
e	2. overuse of land	**b.** resistance to pesticides
f	3. fertile soil	**c.** salinization
b	4. high pesticide use	**d.** malnutrition or starvation
d	5. population pressure	**e.** desertification
a	6. no-till farming	**f.** action of living organisms
c	7. irrigation and evaporation	**g.** loss of topsoil

Concept Mapping

8. Complete the unfinished diagram below to illustrate the connections between the different components.

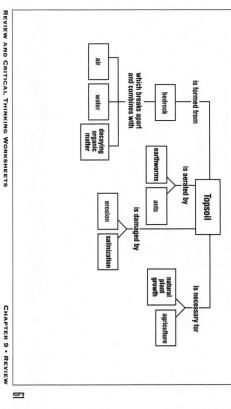

REVIEW AND CRITICAL THINKING WORKSHEETS   CHAPTER 9 • REVIEW **61**

REVIEW AND CRITICAL THINKING WORKSHEETS   **CHAPTERS 8, 9 • ANSWER KEY**

139

CHAPTER 9 REVIEW, CONTINUED

Multiple Choice   In the space provided, write the letter of the word or statement that best answers the question or completes the sentence.

c 1. The biggest loss of arable land is caused by
a. erosion due to floods and droughts.
b. salinization and desertification.
c. use of farmland for urban development.
d. climatic changes.

a 2. The living organisms in fertile soil are found in
a. the surface litter and topsoil.
b. the leaching zone.
c. the subsoil.
d. the bedrock.

b 3. Organic farmers use all of the following EXCEPT
a. compost and animal manure.
b. pesticides such as malathion.
c. crop rotation.
d. cover crops.

d 4. The development of pesticide resistance is an example of
a. malnutrition.
b. persistence.
c. pest control.
d. evolution.

b 5. The green revolution depended on
a. new biodegradable pesticides.
b. high-yielding grain varieties.
c. clearing forest for crop land.
d. powerful fertilizers.

a 6. Erosion is a danger whenever the soil is
a. bare and exposed to wind and rain.
b. plowed along the contour of the land.
c. covered with grass.
d. planted to forest.

b 7. Biological pest control aims to do all the following EXCEPT
a. maintain tolerable pest levels.
b. reduce all insects to low levels.
c. leave non-pest species unharmed.
d. boost plants' natural defenses.

d 8. Salinization may be caused by
a. a rise in groundwater levels.
b. long-term irrigation.
c. salt-tolerant crops.
d. Both a and b are correct.

CHAPTER 9 REVIEW, CONTINUED

Short Answer   Write the answers to the following questions in the spaces provided.

1. Explain how biological insect control can kill only the target pest while chemical insecticides kill many different kinds of insects.

Chemical pesticides are poisons that kill any insect that comes in contact with them. Biological pest control might sterilize the males, broadcast pheromones, or spread bacteria that kill the caterpillars. These biological control methods will therefore not affect non-targeted species.

2. Explain why heavy use of chemical insecticides means that we have to keep developing new ones to use.

Even if a few individuals are not harmed by the chemicals, they will grow and reproduce into a new population that may be resistant. Then we must develop new chemicals to kill the newly-resistant population.

3. Explain how political problems can be more important than agricultural yields in determining whether people go hungry.

When there is a war between countries or within a country, transportation is disrupted. Without transportation, food sent as aid sits in storage rather than being distributed to the people who need it. Also, political struggles are power struggles, and members of one side often withhold food from members of the other side simply to gain an advantage in the power struggle.

4. Describe two farming practices that can help reduce erosion from water.

Mixing organic matter (stems and roots of previous crops, crops planted to be turned back into the ground, compost, or manure) into the soil helps loosen topsoil so water can soak in and not run off. It also feeds earthworms, which open the soil, allowing water to drain into it. Plowing along the contour and planting contour strips with grass helps trap any water running downhill.

CRITICAL THINKING WORKSHEET

AGREE OR DISAGREE

CHAPTER 9

Agree or disagree with the following statements, and support your answer.

1. If we develop salt-tolerant plants, we may be able to use ocean water for crop irrigation.

   Accept any reasonable response. Sample answer: Disagree; using ocean water for irrigation could lead to extreme salinization unless we used huge amounts of water to keep washing the extra salt away. It might work along the coast. However, using large amounts of water to wash away salts might cause soil erosion.

2. The reason the world's deserts are expanding is that the greenhouse effect is warming the climate.

   Disagree; a warmer climate may pose a problem in the future, but deserts are presently expanding from overuse of the areas around them. The plants and soil in these areas are not being given enough time to recover between cropping or grazing.

3. It would be better to apply genetic engineering to develop resistant crops than to rely on chemical pesticides.

   Accept any thoughtful response. Sample answer: Agree; genetic engineering could transfer genes for insect resistance from one plant species into another. Genes from resistant plants that we already eat are almost certainly safer than chemical pesticides. The problem with foreign genes would be in transferring genes causing allergic reactions for some people. Perhaps we need to label food produced by genetic engineering. On the other hand, food is not labeled now to say what pesticides have been used on it.

4. The green revolution made it possible for subsistence farmers to grow more and produce a surplus to sell.

   Disagree; subsistence farmers do not gain from planting the new, high-yielding varieties, because they can't afford the fertilizer, pesticides, and irrigation. For subsistence farmers, the old varieties that do not need the expensive input grow better than the new.

CRITICAL THINKING WORKSHEET

REFINING CONCEPTS

CHAPTER 9

The statements below challenge you to refine your understanding of concepts covered in the chapter. Think carefully, and answer the questions that follow.

1. Clearing land for crops and the resulting destruction of habitat is leading to extinction of plants and animals at a high rate. Why do plant breeders and genetic engineers need to have as broad a base of "wild" plants as possible?

   Our present crop plants were selected and bred from "wild" plants. Breeders and geneticists can only shift genes around. They can't make genes. "Wild" plants are their only source of new genetic material.

2. Organic farming adds material to the litter layer. How does this affect the soil profile?

   Earthworms, beetles, fungi, and bacteria living in the litter layer use organic matter for food. As they decompose the organic matter, it adds to the topsoil layer, and the topsoil grows deeper. The acids from the decomposition leach down and help break up the bedrock, which then adds to the subsoil.

3. People have been farming for the past 10,000 years. If erosion had been going on at the current rate for all that time, we would have reached bedrock long ago. Why has erosion become a serious problem only recently?

   For most of the 10,000 years, farmers did not leave bare soil exposed in huge fields. Small fields and gardens meant less runoff. Crop residue was returned to the fields. Farmers also widely practiced crop rotation, and they let the land recover by letting fields lie fallow every few seasons. Erosion has become a big problem with the widespread use of modern, efficient machinery and farming practices.

CRITICAL THINKING WORKSHEET

INTERPRETING DATA

Read the following information, and answer the questions that follow.

Rice, corn, millet, and wheat are four of the most widely eaten grains. In the raw, these grains each provide 330 Calories or more of energy per 100 g of grain. However, cooking greatly reduces the available energy, especially in rice and millet. Cooked rice provides only about 110 Cal/100 g, and cooked millet about 120 Cal/100 g.

1. Refer to Figure 9-4, on page 229.

a. Calculate the amount of grain produced per person per day in 1995.

300 kg/person/yr divided by 365 days/yr = 0.82 kg/person/day

b. Assume that the average amount of energy provided by cooked grains is 200 Cal/100 g. Calculate how many Calories/person/day were available from grain produced in 1995 assuming that we ate all of the grain we produced.

200 Cal/100 g × 1,000g/kg = 2,000 Cal/kg

2,000 Cal/100 g × 0.86 kg/person/day = 1,720 Cal/person/day.

c. Adults require between 2,000 and 3,500 Cal/day, while children usually require less. Assume that the average person requires 1,800 Cal/day to remain healthy. Assuming that we produced enough grain to feed the people of the world in 1995. Note that this is an oversimplification because humans need a varied diet to stay healthy.

1,720 Cal/day is less than 1,800 Cal/day. Assuming the average person requires 1,800 Cal/day, grain production in 1995 was insufficient to feed the human population.

2. Refer to Figure 9-7, on page 231. What percentage of worldwide arable land was lost between 1985 and 2000?

1.5 billion ha × 1000 mil ha/bil ha = 1,500 million ha of land remaining

1,500 mil ha + 150 mil ha + 50 mil ha + 60 mil ha + 25 mil ha = 1,785 mil ha

(150 + 50 + 60 + 25) / 1,785 = 16% of worldwide arable land

CRITICAL THINKING WORKSHEET

READING COMPREHENSION AND ANALYSIS I

Read the following passage, and answer the questions that follow.

There are two quite simple reasons why I believe that the world will continue to depend on farming, supplemented in an important way by fishing, for most of its food supply in the decades ahead. The first is that the principal food of the world (eaten directly or through animal products), the grains, is relatively low in cost... The second reason is the enormous *weight* of food that is produced each year. In order to supply the calories required for life, it appears that there is little possibility of reducing the dry weight involved. U.S. grain production, measured by *weight*, is more than 15 times greater than steel and 10 times greater than automobiles. The current low cost of food grains — the major primary or secondary source of calories for all people — combined with the enormous volume or weight involved make it most unlikely that much progress will be made in the next fifty years in replacing agriculture by factories.

D. Gale Johnson*

1. What is the source of carbon in most crops today, and what would probably be the carbon source in factory-made crops?

The carbon in crops today comes from the carbon dioxide in air, while carbon in factory-made crops would probably come from a fossil fuel.

2. How does the source of energy to make today's crops differ from the possible sources of energy we would need to make crops in factories?

Crop plants efficiently get their energy from sunlight. Factory-made food would probably get its energy from fossil, nuclear, or hydroelectric fuels, since those are the most cost-effective fuels available today.

3. The author wrote this article more than thirty years ago. Do you think he was right or wrong?

Accept any thoughtful answer. Sample answer: He was probably right. Crop plants do "for free" what would be very expensive and wasteful if done artificially.

*From "Food: The World's People Won't Go Hungry" by **D. Gale Johnson** from *Toward the Year 2018*, edited by the Foreign Policy Association. Copyright © 1968 by the Foreign Policy Association. Reprinted by permission of **the author and the Foreign Policy Association.**

Name _____ Class _____ Date _____

CHAPTER 9

READING COMPREHENSION AND ANALYSIS II

Read the following passage, and answer the questions that follow.

Under primitive agricultural conditions the farmer had few insect problems. These arose with the intensification of agriculture—the devotion of immense acreages to a single crop. Such a system set the stage for explosive increases in specific insect populations. Single-crop farming does not take advantage of the principles by which nature works; it is agriculture as an engineer might conceive it to be.

Rachel Carson*

1. Why would planting large areas with a single type of crop plant "set the stage" for large increases in insect populations?

Planting large areas with one type of crop plant gives pests of that crop a one-stop food bonanza. In such a system, pests are not confronted with the usual heterogeneity characteristic of natural landscapes. Single-crop systems also do not provide refuges for the predators that otherwise keep pest populations under control.

2. How does single-crop farming, also known as monocropping, "not take advantage of the principles by which nature works?"

Single-crop farming does not provide the checks and balances that normally regulate organism's population sizes under natural conditions. Natural habitats are variable, with food sources being thinly scattered over large areas, and predators occupying the same habitat as their prey. Monocropped fields do not provide this variability or these obstacles, so pests are able to use much more of their energy for producing young, and more of these young survive to reproduce.

*From *Silent Spring* by Rachel Carson. Copyright © 1962 by Rachel L. Carson; copyright renewed © 1990 by Roger Christie. All rights reserved. Reprinted by permission of *Houghton Mifflin Company.*

88

CHAPTER 9 • CRITICAL THINKING

HOLT ENVIRONMENTAL SCIENCE

Name _____ Class _____ Date _____

CHAPTER 10

BIODIVERSITY

Matching Match each example in the left column with the appropriate term from the right column.

b 1. a species not native to a particular region   a. keystone species

e 2. any species that is likely to become endangered if they are not protected   b. exotic species

a 3. species that are very important to the functioning of an ecosystem   c. extinct

d 4. any species whose numbers have fallen so low that it is likely to become extinct in the near future if not protected immediately   d. endangered species

c 5. status of a species when the very last individual dies   e. threatened species

Now match each definition in the left column with the vocabulary word in the right column.

d 6. Florida panther   a. keystone species

c 7. American passenger pigeon   b. exotic species

a 8. sea otter   c. extinct species

e 9. northern spotted owls   d. endangered species

b 10. melaleuca trees   e. threatened species

89

REVIEW AND CRITICAL THINKING WORKSHEETS

CHAPTER 10 • REVIEW

Name _____ Class _____ Date _____

11. In each blank circle, write the name of the organism from the list that corresponds to the specific threat in the attached box. Then draw a line from each of the specific threats to the corresponding general threat in the middle of the diagram.

Concept Mapping

Concept map elements:

- African elephant
- American bald eagle
- the ivory trade
- thinning eggshells threatened its survival
- Florida panther
- clearing of forests in the American Southeast
- melaleuca tree
- Native Florida plants
- pollution
- poaching
- habitat destruction
- exotic species
- hunting
- Snail darter
- Tellico Dam project
- Sea otter
- Pacific Coast fur trade
- killed for their tongues and hides
- draining of wetlands displaced this bird
- Bison
- Whooping Crane

List of Organisms:

African elephant · American bald eagle · Bison · Florida panther · Native Florida Plants · Sea otter · Snail darter · Whooping crane

---

Name _____ Class _____ Date _____

Short Answer   Write the answers to the following questions in the spaces provided.

1. Explain why many conservationists now concentrate on protecting entire ecosystems rather than individual species.

By concentrating on saving entire ecosystems, we may be able to save most of the species in an ecosystem rather than just the ones that are endangered or threatened. Also, the health of the entire biosphere depends upon preserving each of the individual ecosystems that comprise it.

2. Explain the difference between an *endangered* species and a *threatened* species.

An endangered species is a species whose numbers have fallen so low that it is likely to become extinct in the near future if it is not protected soon. A threatened species is a species that is likely to become endangered if it is not protected soon. Unlike an endangered species, a threatened species' population has not yet to become so low that the species faces extinction.

3. Briefly explain three ways to save individual species.

One method for saving species is through captive-breeding programs, in which animals are bred and their populations managed in zoos and animal parks. A second way to save species is botanical gardens, which preserve live plant and insect species. A third method of saving individual species is through germ-plasm banks, where the reproductive (germ) cells of species are stored for the future. Here, plants may be stored as seeds, and animals may be stored as frozen sperm and eggs.

4. Why do many scientists believe that the Earth is currently experiencing a mass extinction?

Many scientists fear that the Earth is experiencing a mass extinction because the rapidly growing human population is putting new stress on organisms and their habitats. Biologically diverse natural areas, such as tropical rain forests, are being cleared to provide for a rapidly growing human population. At the same time, humans have caused many extinctions by introducing nonnative species and by hunting.

CRITICAL THINKING WORKSHEET

ANALYZING ARGUMENTS

The gray wolf is a species that people have maligned in campfire tales and stories such as *Little Red Riding Hood*. Recently, the gray wolf was reintroduced into Yellowstone National Park. Shortly thereafter, a federal judge ruled that the reintroduction program was illegal for various reasons.

1. Why is reintroducing the gray wolf important for the Yellowstone ecosystem?

The gray wolf is important because, as a predator, it helps to maintain the balance of nature in this ecosystem, keeping the population of large and small mammalian herbivores in check. Therefore, preserving the gray wolf helps preserve the entire ecosystem.

2. Nearby ranchers claim that the wolves will prey on their livestock. Is it possible to evaluate this claim "before the fact?"

It is possible to study the gray wolf in its habitat to document its diet. One could also review past records kept by ranchers. Ultimately, though, there is probably no way to know beforehand whether one or more wolf populations will begin preying on livestock. (Note: Studies have shown that there is little evidence to support the claim that wolves are responsible for killing large numbers of sheep or cattle.)

3. Is it possible to predict the effect(s) of eliminating a species from an ecosystem? Explain your answer.

No; we do not fully understand what effects we are having on the environment when we eliminate a species from an ecosystem. However, we have seen many cases where the results have been negative. Such tampering has been at least partly responsible for losses of populations and species we have witnessed in the last several centuries we now have.

CRITICAL THINKING WORKSHEET

AGREE OR DISAGREE

Agree or disagree with the following statements, and support your answer.

1. When battles between developers and environmentalists are worked out, usually neither side gets everything they want, but both sides get something. This compromising approach is effective enough to save endangered species.

Accept any thoughtful answer. Sample answers: Disagree; endangered species are already in a compromised position by being at the brink of extinction and further compromising would almost certainly result in their extinction. Agree: Human needs always take priority. We can preserve these species through captive-breeding programs at zoos or in germ-plasm banks.

2. Do you think the Biodiversity Treaty should have been signed?

Accept any thoughtful answer. Sample answers: Agree: money given to poorer countries for the protection of potentially valuable species would benefit every country. President Clinton was right in signing the treaty. Disagree: The treaty should have stated exactly how the money would be spent by the poorer countries before the Biodiversity Treaty was signed. President Bush acted correctly.

3. To protect biodiversity worldwide, many conservationists suggest that at least 10 percent of the Earth's land be set aside as protected preserves. This percentage is a good "hard and fast" rule to be used in all cases.

Accept any reasonable response. Sample answer: Disagree: some species may need more than 10 percent of an area for their species to survive. Also , it is possible that saving only 10 percent of a biodiversity "hot spot" may not be enough if the surrounding areas become overstressed, such as from drought or pest over-population.

CRITICAL THINKING WORKSHEET

INTERPRETING EVIDENCE

CHAPTER 10

Name _____ Class _____ Date _____

Write the answers to the following questions in the spaces provided.

1. Migratory species, such as many birds, salmon, and the monarch butterfly, can be especially vulnerable to habitat loss. At the same time, they can be instrumental in preserving ecosystems. Explain why both of these statements are true.

Migratory species can be especially vulnerable because either or both of their environmental destinations (generally both breeding grounds and wintering grounds), can be threatened or destroyed by human activities. In addition, they can encounter a variety of difficulties along their migratory routes, especially if they need to rest each night along the way. Yet by making an effort to save migratory species, we would also be saving a variety of ecosystems in the United States and in other countries in the Western Hemisphere.

2. What conditions do the Florida panthers need in order to come back from the brink of extinction? Why is this unlikely to occur?

The Florida panther requires a large area of forest or swampland for feeding and breeding. Today, however, much of the panther's habitat has been broken up into small, open areas by fences, roads, and canals. Destruction of its habitat has made the Florida panther an endangered species. Fear of the predatory panther also makes it unlikely that humans will save this species.

3. In the summer of 1998, numerous fires spread through parts of Florida. Explain how importing melaleuca trees in the early 1900s may have contributed to this situation.

In the early 1900s, real estate developers hoped that planting the exotic melaleuca trees, originally from Australia, would dry up the Everglade wetlands and make them suitable for development. The trees did help dry up the land. The drier land probably contributed to the tinderbox conditions that have resulted in the wildfires in Florida.

CRITICAL THINKING WORKSHEET

READING COMPREHENSION AND ANALYSIS I

CHAPTER 10

Name _____ Class _____ Date _____

Read the following passage, and answer the questions that follow.

The immense diversity of the insects and flowering plants combined is no accident. The two empires are united by intricate symbioses. The insects consume every anatomical part of the plants, while dwelling on them in every nook and cranny. A large fraction of the plant species depend on insects for pollination and reproduction. Ultimately, they owe them their very lives, because insects turn the very soil around their roots and decompose dead tissues into the nutrients required for continued growth.

So important are insects and other land-dwelling arthropods that if all were to disappear, humanity probably could not last more than a few months.

Edward O. Wilson*

1. What type of species interaction is the author referring to (hint: see Chapter 2), and what does it imply about the evolution of the Earth's biodiversity?

The author is referring to coevolution, the process by which species evolve in response to each other. His point about the coevolution of insects and flowering plants implies that, since these species make up the majority of the species on Earth, the biodiversity we have has been primarily driven by coevolution.

2. What is the reasoning behind the author's last statement?

If insects and other land-dwelling arthropods disappeared, flowering plants that rely on insects for pollination would become extinct. For these and other reasons, our food supply would be threatened. Without plant-based foods, humans would not exist for very long either.

3. Popular images of biodiversity often focus on large endangered mammals, such as pandas and tigers. Why do you think the actual situation is not as well-known?

It is more difficult to capture public attention with pictures of beetles and flowering plants than it is with pictures of cute, fuzzy creatures. The media tends to focus on such images, fueling misperceptions about where the greatest threats to biodiversity lie.

*From *The Diversity of Life* by Edward O. Wilson. Copyright © 1992 by Edward O. Wilson. Published by **Harvard University Press**, Cambridge, Mass. Reprinted by permission of the publisher.

Name _____ Class _____ Date _____

CRITICAL THINKING WORKSHEET

READING COMPREHENSION AND ANALYSIS II

CHAPTER 10

Read the following passage, and answer the questions that follow.

Coffee is traditionally grown under a canopy of shade trees. Because of the structural and floristic complexity of the shade trees, traditional coffee plantations have relatively high biodiversity. However, coffee plantations increasingly are being transformed into industrial plantations with little or no shade.... The way that coffee production evolves in the coming decades is likely to have a tremendous impact on its ability to provide a refuge for tropical biodiversity.

Ivette Perfecto et al.*

1. The author states that traditional coffee plantations have relatively high biodiversity. What does she mean by this statement?

   The author means that traditional coffee plantations tend to be inhabited by a large number of different species of organisms. In order to support these organisms, traditional plantations must provide a variety of potential niches that different species can exploit.

2. The author offers an explanation for the biodiversity of traditional coffee plantations. What does this explanation imply about the biodiversity of tropical rain forests?

   The author states that traditional coffee plantations have high biodiversity because the shade trees that grow there have complex structures and flowers. This implies that the high biodiversity of tropical rain forests might be due to the complexity of the vegetation that grows there.

3. What point is the author making in her last sentence?

   The author is pointing out that given the difference in biodiversity between traditional and industrial coffee plantations, the extent to which plantations change overall will greatly affect their ability to harbor a diverse assemblage of species. She is also implying that there will be a need for biodiversity refuges in coffee-growing regions—in other words, that much of the natural habitat will be destroyed.

76

*From "Shade Coffee: A Disappearing Refuge for Biodiversity" by Ivette Perfecto et al. from *BioScience*, vol. 46, no. 8, September 1996, p. 598. Copyright © 1996 by the **American Institute of Biological Sciences.** Reprinted by permission of the publisher.

CHAPTER 10 • CRITICAL THINKING

HOLT ENVIRONMENTAL SCIENCE

Name _____ Class _____ Date _____

CHAPTER REVIEW

ENERGY

CHAPTER 11

Matching Match each example in the left column with the appropriate term from the right column.

_b_ 1. biomass, sun, wind, water   **a.** fossil fuels

_e_ 2. fission and fusion   **b.** renewable resources

_a_ 3. oil, natural gas, and coal   **c.** electric generator

_c_ 4. magnetic fields and rotating turbines   **d.** biomass

_d_ 5. sugar cane, litter from chicken coops, and wood   **e.** nuclear energy

Concept Mapping

6. Demonstrate how energy is converted using the diagram below. Draw lines connecting each energy source through the complete process of energy conversion. Each energy source may be converted through more than one process.

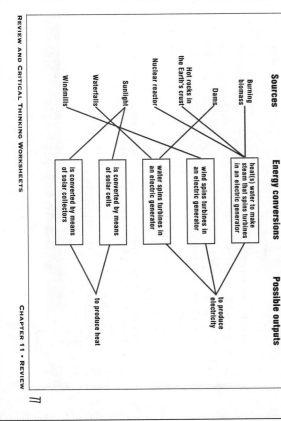

Sources	Energy conversions	Possible outputs
Burning biomass	heat(s) water to make steam that spins turbines in an electric generator	
Nuclear reactor	wind spins turbines in an electric generator	to produce electricity
Hot rocks in the Earth's crust	water spins turbines in an electric generator	
Dams	is converted by means of solar cells	
Sunlight	is converted by means of solar collectors	to produce heat
Waterfalls		
Windmills		

REVIEW AND CRITICAL THINKING WORKSHEETS

CHAPTER 11 • REVIEW

7

147

CHAPTER 11 REVIEW, CONTINUED

Name _____ Class _____ Date _____

Short Answer Write the answers to the following questions in the spaces provided.

1. Distinguish between renewable and nonrenewable resources. Provide a specific example of each.

   Resources such as fossil fuels that are being used more quickly than they can be replaced are considered to be nonrenewable. Renewable resources are continually produced, and resources such as wind and sunlight are considered to be inexhaustible.

2. How is a nuclear power plant similar to a plant that burns fossil fuels? How are they different?

   Both plants generate heat, converting water into steam. This steam rotates the turbines of an electric generator. However, the means by which the heat is generated is different. Nuclear power plants generate heat by means of radioactive decay. As radioactive materials decay into a less energetic state, they release a lot of energy, heating the water. Burning fossil fuels heats water simply by fire.

3. Define energy conservation, and give three examples not listed in the book.

   Energy conservation means using materials and energy wisely. Additional methods of conserving energy include hanging clothes out to dry, washing hands with cold instead of heated water, and using hand tools instead of electric or gas-powered tools.

4. Is wind energy a form of solar energy? Justify your answer.

   Yes, wind energy is an indirect form of solar energy. Wind is the movement of air caused by the uneven heating of the Earth's surface by the sun.

CRITICAL THINKING WORKSHEET

ANALOGIES

CHAPTER 11

Name _____ Class _____ Date _____

Mark the letter of the pair of terms that best completes the analogy shown. An analogy is a relationship between two pairs of words or phrases written as a:b::c:d. The symbol : is read *is to*, and the symbol :: is read *as*.

Example

keyboard : type ::
   a. plane : land
   b. dog : eat
   √ c. scissors : cut
   d. rock : hard
   e. grass : green

1. Electric generator : electricity ::
   a. heater : blanket
   b. heating : window
   √ c. freezing : ice
   d. freezer : ice
   e. plants : photosynthesis

2. Power plant : fossil fuels ::
   √ a. nuclear plant : nuclear fuels
   b. power plant : radioactive material
   c. fossil fuel : biomass
   d. resources : water
   e. electricity : wire

3. nonrenewable resource : limited ::
   a. biomass : heat
   b. nuclear energy : nuclear plant
   c. oil : coal
   d. solar energy : disadvantage
   √ e. renewable resource : abundant

4. Fission : splitting ::
   a. chain reaction : neutron
   √ b. fusion : combining
   c. nuclear plant : radioactive waste
   d. geothermal energy : drilling
   e. hydrothermal energy : dam

5. Nuclear fusion : star ::
   a. electricity : electric generator
   √ b. combustion : car
   c. steam : biomass
   d. sunlight : solar cells
   e. nuclear fission : heat

6. solar energy : solar cell ::
   √ a. power plant : fossil fuel
   b. mechanical energy : electric generator
   c. nuclear fusion : nuclear fission
   d. solar energy : solar heating
   e. liquid fuels : biomass

7. Liquid fuel : biomass ::
   a. hydrogen : liquid
   b. oil : coal
   √ c. ethanol : fruit
   d. gasohol : gas
   e. solar cells : sunlight

Name _____ Class _____ Date _____

CHAPTER 11

INTERPRETING OBSERVATIONS

Read the following scenario, and then answer the questions below.

Imagine that you are a civil engineer who was just hired by the government of Iceland to plan a city. On the airplane you read about the "land of fire and ice," and learn that it was formed from tectonic activity. As a result, there are many volcanoes, geysers, and hot springs on the island. Also, because Iceland is located in northern latitudes, it receives a lot of snowfall in winter.

As your plane zooms in for a landing, you notice that the annual springtime snowmelt has formed large lakes and colossal waterfalls—two of which are very impressive! Some volcanoes lining an overcast sky are still capped with snow. You notice that there aren't very many trees around, only grass and shrubs in a rocky landscape.

1. What are two sources of energy you would use for your city? Explain your answer.

I would set up turbines at the base of the waterfalls or create a hydroelectric dam in one of the lakes. I could also use steam from a geyser to spin the turbine of an electric generator.

2. Using materials already at your disposal, how could you construct the houses so that they are well-insulated? How could you heat the water?

I would build houses with thick stone walls to contain heat and use water from the hot springs.

3. Would it be effective to use solar cells or solar collectors on the roofs to provide electricity or hot water? Explain your answer.

No, it would not be effective to use solar cells or solar collectors on the roofs to provide electricity or hot water. The sky is usually overcast, and the northern latitudes do not receive much sunlight.

Name _____ Class _____ Date _____

CHAPTER 11

AGREE OR DISAGREE

Agree or disagree with the following statements, and support your answer.

1. Producing electricity on a large scale inevitably has environmental costs.

Accept any thoughtful answer. Sample answer: Agree; renewable resources, such as wind and sunlight, are not abundant enough in all parts of the world to be able to produce electricity without cost to the environment. All other current energy sources are either nonrenewable or harm the environment.

2. Electricity should cost more money so that people will use less.

Accept any thoughtful response. Sample answers: Agree; people don't want to pay more money for electricity, so they will learn to conserve it. Disagree; people who live in extreme heat or cold climates risk their health if they cannot afford sufficient air conditioning or heating.

3. Fifty years from now, our major source of energy will be sunlight.

Accept any thoughtful response. Samples answers: Agree; if enough solar cells are positioned in places that receive plenty of sunlight, perhaps enough energy could be generated for worldwide consumption, provided that we conserve energy and use it wisely. Disagree; cities or towns located in extreme northern or southern latitudes do not receive enough sunlight to generate enough electricity to be self-sustaining.

Name _____ Class _____ Date _____

CRITICAL THINKING WORKSHEET

INTERPRETING DATA

CHAPTER 11

Examine the data in the table below, and answer the questions that follow.

	How energy is used worldwide	How energy is used in the United States
Electricity generation	19%	29%
Industry	28%	19%
Transportation	16%	25%
Commercial, public, residential	17%	15%
Other	20%	12%

1. Use the data in the table above to compare how energy is used worldwide with how it is used in the United States. What recommendations would you make to city officials in the United States to encourage energy conservation? Support your answer with data from the table above.

The energy used for transportation in the United States is almost twice that of the rest of the world. City officials should plan and implement efficient and inexpensive public transportation systems that rely on clean-burning fuels or electricity from a renewable source of energy.

2. The United States uses 29% of its energy for electricity generation and 19% for industry, while the percentages used worldwide are nearly opposite. Explain what might account for this difference. Hint: consider the two groups into which most countries fall.

Accept any reasonable response. Sample answer: The United States probably uses more energy for electricity than for industry because it is a developed nation. It has already built much of its infrastructure and has a comparatively wealthy populace that can afford electricity-using devices, such as televisions and computers. However, the majority of people worldwide live in developing nations, which tend to focus on industry and whose citizens generally have less money to spend on devices that require electricity.

Name _____ Class _____ Date _____

CRITICAL THINKING WORKSHEET

READING COMPREHENSION AND ANALYSIS I

CHAPTER 11

Read the following passage, and answer the questions that follow.

Consumers are not interested in buying raw energy, but in getting access to energy services—heat, light, hot water, etc. If those services can be provided more cheaply through efficiency than by generating more energy, then it is just good business for utilities to supply those services, and to enable their consumers to use energy more efficiently.

L. Hunter Lovins*

1. How might efficiency lower an energy provider's costs?

As resources become more scarce, energy providers must pay more money for access to them. If an energy provider could develop a more efficient way to use resources, that provider would not have to spend as much money obtaining them.

2. Do you agree with the author's statement that "consumers are not interested in buying raw energy?" Explain your answer.

Accept any thoughtful answer. Sample answer: Yes; consumers do not necessarily associate quantities of energy with the electrical appliances they operate or water they use. Consumers want to continue using products which use energy but have little concern for the total amount of energy used.

3. What problems, if any, do you see with providing energy more efficiently?

Accept any thoughtful answer. Sample answer: Providing energy more efficiently (and more cheaply) may prevent consumers from making the connection between their energy use and the depletion of nonrenewable resources. People may not realize the importance of using resources wisely.

*From "Efficiency: Less Energy, More Power," by L. Hunter Lovins in *Environment*, 2nd edition. Copyright © 1998 by **Saunders College Publishing, a division of Harcourt Brace & Company.** Reprinted by permission of the publisher.

Left Page

Name _____ Class _____ Date _____

READING COMPREHENSION AND ANALYSIS II

CHAPTER 11

Read the following passage, and answer the questions that follow.

Tidal power utilizes the twice daily changes in sea levels to generate electricity. It is suitable only for certain coastlines, but it certainly offers great possibilities in places where the tides vary twenty to a hundred feet per day. Wave power is another dynamic area awaiting development.

Helen Caldicott*

1. What does this passage generally imply about options for providing us with energy?

Accept any reasonable response. Sample answer: this passage implies that any source of natural energy,

such as waves, can be tapped to supply energy to meet human needs.

2. How is the author using the word *dynamic* in this passage?

The author is using dynamic to refer to the potential of wave power for rapid development as an energy

source.

3. As a scientist, how would you investigate the energy-supplying potential of tides?

Accept any thoughtful answer. Sample answer: I would use a pipe placed at low-tide sea level to draw water

as the tide came in. The pressure of the rising water rushing through the pipe could be used to move a turbine

and produce electricity.

*from *If You Love This Planet* by Helen Caldicott. Copyright © 1992 by W. W. Norton & Company, Inc. Reprinted by permission of the publisher.

84

CHAPTER 11 • CRITICAL THINKING

HOLT ENVIRONMENTAL SCIENCE

Right Page

Name _____ Class _____ Date _____

WASTE

CHAPTER 12

Matching   Match each example in the left column with the appropriate term from the right column.

_d_ 1. household rubbish

_c_ 2. landfill water containing hazardous waste

_e_ 3. waste containing lead, mercury, or cadmium

_a_ 4. newspaper, cotton fibers, leather

_b_ 5. polyester, nylon, plastic

_f_ 6. steam, ash, electrostatic precipitator

a. biodegradable materials

b. synthetic materials

c. leachate

d. municipal solid waste

e. hazardous waste

f. incinerator

Concept Mapping

7. Demonstrate your understanding of municipal solid waste management. For each example below, draw a line from each material to the appropriate solid-waste management technique to each possible outcome.

Materials

plastic utensils

paper bag

leaves

batteries

Waste management techniques

recycling

incinerator

landfill

compost

Possible outcomes

rich soil

combustible gases such as methane

leachate seepage into groundwater

air pollution

debris goes to landfill

reused as another material

REVIEW AND CRITICAL THINKING WORKSHEETS

85

CHAPTER 12 • REVIEW

Name _____ Class _____ Date _____

CHAPTER 12 REVIEW, CONTINUED

Multiple Choice In the space provided, write the letter of the word or statement that best answers the question or completes the sentence.

__b__ 1. Solid waste includes all of the following EXCEPT
   a. agricultural waste.
   b. methane.
   c. plastics.
   d. food waste.

__b__ 2. The ash produced by incinerators is _____ than other solid waste.
   a. less toxic
   b. more toxic
   c. as toxic
   d. more recyclable

__a__ 3. The law that makes owners of hazardous-waste sites responsible for cleanup is
   a. the Superfund Act.
   b. the Incinerator Act.
   c. the EPA.
   d. the Love Canal Act.

__c__ 4. Surface impoundment relies on which physical process?
   a. decontamination
   b. compression
   c. evaporation
   d. interjection

__d__ 5. Discarded material that is not in liquid or gas form is technically called
   a. crud.
   b. junk.
   c. sludge.
   d. solid waste.

__b__ 6. The amount of yard waste produced in the United States in 1995 was about
   a. 45,000 tons.
   b. 20,000 tons.
   c. 10,000 tons.
   d. 30,000 tons.

__b__ 7. The best way to make sure radioactive waste does not pose a serious threat to humans is to
   a. treat it with chemicals.
   b. store it where it can decay far away from people and water sources.
   c. deep-well inject it.
   d. decontaminate it.

__d__ 8. The amount of municipal solid waste going to landfills could be reduced by
   a. composting yard waste.
   b. recycling.
   c. reusing products.
   d. All of these answers are correct.

Name _____ Class _____ Date _____

CHAPTER 12 REVIEW, CONTINUED

Short Answer Write the answers to the following questions in the spaces provided.

1. Define recycling, and provide two examples.

   Recycling involves breaking down a product or material, such as a glass bottle or newspaper, to use as the raw material to make another product. One example of recycling is melting down old aluminum cans to use to make new ones. Another example of recycling is breaking wood waste into chips and pressing the chips into particle board for constructing furniture.

2. Explain how plastic bags can be labeled "biodegradable" when plastic is nonbiodegradable.

   The reason for this apparent contradiction is that biodegradable plastic bags are made of more than just plastic; the nonplastic portion is biodegradable. Plastics are composed of molecules linked together in very stable chains. Manufacturers of biodegradable plastics weaken the chains by adding cornstarch or other organic compounds to the plastic. In bags made with cornstarch, bacteria eat the cornstarch, thinning and weakening the bag so that it eventually degrades into small pieces of plastic.

3. Explain why some types of waste are considered hazardous waste, and provide examples.

   Some types of waste are considered hazardous waste because they are toxic, highly corrosive, or highly flammable, or because they explode easily. Examples of hazardous wastes include certain solvents, lead, mercury, and battery acid.

CRITICAL THINKING WORKSHEET — CHAPTER 12

Name _____ Class _____ Date _____

ANALOGIES

Mark the letter of the pair of terms that best completes the analogy shown. An analogy is a relationship between two pairs of words or phrases written as a:b::c:d. The symbol : is read *is to*, and the symbol :: is read *as*:

Example
keyboard : type ::
___ a. plane : land
___ b. dog : eat
_✓_ c. scissors : cut
___ d. rock : hard
___ e. grass : green

1. solid waste : leachate ::
   ___ a. tree : sap
   ___ b. pie : filling
   ___ c. liquid waste : toxins
   _✓_ d. body : sweat
   ___ e. salty soil : salty water

2. Oil : recycling ::
   ___ a. battery : community collection site
   ___ b. solid waste : hazardous waste
   ___ c. compost : leaves
   _✓_ d. waste reduction : material conversion
   ___ e. compost : nylon

3. Pond : surface impoundment ::
   ___ a. deep-well injection : drill
   _✓_ b. baby : diapered baby
   ___ c. oil : groundwater
   ___ d. hazardous waste : incineration
   ___ e. degradable : biodegradable

4. Leachate : groundwater ::
   ___ a. synthetic material : compost
   ___ b. energy : heat
   ___ c. plastic : landfill
   _✓_ d. oil : drilling
   ___ e. ash : air

5. Hazardous waste: deep-well injection ::
   _✓_ a. solid waste : landfill
   ___ b. solid waste : waste management
   ___ c. leachate : landfill
   ___ d. surface impoundment : compost
   ___ e. biodegradable material : heat

6. radioactive : hazardous ::
   ___ a. leachate : impoundment
   _✓_ b. biodegradable : solid
   ___ c. bottles : plastic
   ___ d. compost : hazardous
   ___ e. solid : waste

7. Cotton: biodegradable ::
   ___ a. incinerator : ash
   ___ b. yard waste : nonbiodegradable
   _✓_ c. polyester : nonbiodegradable
   ___ d. over-packaged : not packaged
   ___ e. bomb : flammable

CRITICAL THINKING WORKSHEET — CHAPTER 12

Name _____ Class _____ Date _____

THINKING ABOUT PROCESSES

Read the following scenario, and answer the questions that follow.

In an ideal world, tires sent for recycling would be ground into small particles and reformed into new tires. But it is actually not that simple. Tires are made through a process, called *vulcanization*, in which raw rubber is heated and combined with sulfur to make a sturdy, elastic compound. Without the addition of sulfur, rubber would have the consistency of clay. However, the sulfurized rubber is so chemically stable that chemical bonds cannot form between tire particles and added rubber, so old tires cannot be easily converted into new ones. Currently, the only widely-used technology for recycling tires involves processing them with dangerous chemicals, such as chlorine and sulfur dioxide.

1. Based on what you have learned in this chapter, what might be some of the disadvantages of processing tire rubber with chemicals?

   Processing rubber with dangerous chemicals will probably create hazardous waste. It is difficult to store liquid hazardous waste so it doesn't eventually seep into groundwater and contaminate water supplies.

2. We rely heavily on microbes to decompose organic wastes in landfills and compost heaps. There is one type of bacteria, *Sulfolobus acidocaldarius*, that thrives only on sulfur in Yellowstone's hot springs. How might this bacteria help us recycle tires?

   Sulfolobus bacteria could eat the sulfur in tires, leaving behind rubber. We could then use that rubber, in combination with sulfur, to make new tires.

3. Scientists use the term *biodesulfurization* to describe the process explored in question 2. Based on the words in this term and on the scenario above, what do you think biodesulfurization means?

   Biodesulfurization means the process of (-ization) using living organisms (bio-) to remove (-de-) the element sulfur (-sulfur-) from a material.

Worksheet 1 (page 90)

Name _____ Class _____ Date _____

AGREE OR DISAGREE

CHAPTER 12

Agree or disagree with the following statements, and support your answer.

1. Leachate is a form of hazardous waste.

   Agree. leachate is basically a hazardous chemical "soup," and therefore should be treated like other forms of liquid hazardous waste.

2. Compost is biodegradable.

   Accept any reasonable answer. Sample answers: Disagree: the term "biodegradable" implies that something is capable of decomposing. Technically, compost is the brown, crumbly material made from already decomposed vegetable and animal materials. Agree: compost heaps decompose organic "material," and until the heap is entirely decomposed it will contain biodegradable material.

3. There is little danger of hazardous waste entering groundwater if the waste is disposed of through deep-well injection.

   Accept any reasonable response. Sample answers: Disagree: geologic forces could cause the rock layers to move, possibly cause the hazardous waste to migrate upward to the level of groundwater. Agree: if there is no danger of geologic activity in the area, responsible deep-well injection places the hazardous waste too far below groundwater level for it to contaminate groundwater.

90

CHAPTER 12 • CRITICAL THINKING

HOLT ENVIRONMENTAL SCIENCE

Worksheet 2 (page 91)

Name _____ Class _____ Date _____

READING COMPREHENSION AND ANALYSIS

CHAPTER 12

Read the following passage, and answer the questions that follow. Also read page 423 of the textbook for an explanation of the role of economics in environmental decision making, including the decision of whether or not to pollute.

The rational man finds that his share of the cost of the wastes he discharges into the commons is less than the cost of purifying his wastes before releasing them. Since this is true for everyone, we are locked into a system of "fouling our own nest," so long as we behave only as independent, rational, free-enterprisers.

Garrett Hardin*

1. What idea is the author trying convey? Hint: The *commons* means resources that we all share or hold in common, such as clean air, water, and soil.

   The author is saying that a person acting only out of self-interest will tend to pollute because it is cheaper to pollute than clean up one's own pollution. Since everyone is faced with this same choice, we will continue to pollute the environment ("our nest") so long as people act solely out of self-interest.

2. What does this passage imply about the difference between independent and collective rational behavior?

   The passage implies that behavior that is rational for the individual is not necessarily rational for the society in which the individual lives.

3. How can we as a society prevent "fouling our nest?"

   We could award tax incentives to businesses and individuals who produce and buy products that help keep our air, water, and soil clean. We could also avoid fouling our environment by imposing further restrictions on individuals and businesses for the good of the whole. Perhaps the best way to prevent fouling the environment is to modify our behavior so that we use materials and energy as efficiently as possible. For example, businesses are increasingly finding that taking into account the environmental costs of their actions saves them money in the long run.

*From "The Tragedy of the Commons" by Garrett Hardin from *Science*, vol. 162, December 13, 1968, pp. 1243–1268. Copyright © 1968 by the **American Association for the Advancement of Science**. Reprinted by permission of the publisher.

REVIEW AND CRITICAL THINKING WORKSHEETS

CHAPTER 12 • CRITICAL THINKING

91

Copyright © by Holt, Rinehart and Winston. All rights reserved.

Name _____ Class _____ Date _____

READING COMPREHENSION AND ANALYSIS II

CHAPTER 12

Read the following passage, and answer the questions that follow.

You don't get these types of incinerators and chemical plants being compatible with clean industries or office towers. To create white-collar office jobs you have to attract the population, and usually people like to live near where they work.

When the toxic landfill came into Warren County, North Carolina, giving birth to the environmental justice movement, the county started to lose major businesses, because people started to identify the county with hazardous wastes.

Robert Bullard*

1. What happens when a community that produces hazardous wastes refuses to store them? **The community's hazardous waste must be transported to another community. The community which stores the waste becomes devalued, and the community that produces the waste is not forced to confront the negative effects of their waste.**

2. Do you agree with the author's idea that incinerators and chemical plants are not "compatible" with clean industries or office towers? Why? **Accept any thoughtful answer. Sample answer: Yes; people do not want their communities associated with hazardous-waste sites. Consequently, less "industrial" industries and businesses that tend to work in office towers would try to avoid this negative association.**

3. Imagine that a hazardous-waste producer in a neighbor state has proposed storing its waste in your community. How do you think your community would react? **Accept any thoughtful answer. Sample answer: Our community would approve this plan if enough money was offered because our area already has several waste-producing industries that deter clean industries from locating here.**

*From "Some People Don't Have 'The Completion for Protection'" by Robert Bullard from *E: The Environmental Magazine*, vol. IX, no. 4, July/August 1998. Copyright © 1998 by *Earth Action Network*. Reprinted by permission of the publisher.

Name _____ Class _____ Date _____

POPULATION GROWTH

CHAPTER 13

Matching Match each example in the left column with the appropriate term in the right column.

d 1. maximum growth rate of a population   a. limiting resources

f 2. effect of factors that limit population growth   b. hunter-gatherers

c 3. movement of organisms out of a population   c. emigration

a 4. factors that can limit population growth   d. biotic potential

b 5. people that often change locations to find food   e. agricultural revolution

e 6. transition from hunting and gathering to cultivating plants and raising animals   f. environmental resistance

Graphing

7. A population of rabbits moves to a new habitat with a high carrying capacity. Several generations later a severe drought reduces the carrying capacity of the land. Complete the population growth curve below to show how the rabbit population responds to these changes.

Old carrying capacity

New carrying capacity

Population Growth Curve

Name _____ Class _____ Date _____

CHAPTER 13 REVIEW, CONTINUED

Multiple Choice In the space provided, write the letter of the word or statement that best answers the question or completes the sentence.

a 1. The graph of human population growth over time since 1200 A.D. looks like
  a. a J-curve.
  b. an S-curve.
  c. a horizontal line.
  d. a straight 45° line.

b 2. Plant populations are limited by all of the following EXCEPT
  a. water.
  b. nesting sites.
  c. mineral nutrients.
  d. light.

c 3. A population will shrink if deaths + emigrants exceed
  a. deaths + births.
  b. immigration − emigration.
  c. births + immigrants.
  d. the carrying capacity.

c 4. Developed countries tend to have relatively even numbers
  a. of limiting resources.
  b. of economic opportunities.
  c. of people in each age group.
  d. urban problems.

d 5. Environmental resistance might decrease with the following:
  a. a higher biotic potential
  b. a larger population
  c. more immigrants
  d. more food

b 6. About how long did it take the human population to double from 2 billion to 4 billion people?
  a. 130 years
  b. 45 years
  c. 95 years
  d. 175 years

a 7. A population's biotic potential represents its growth rate
  a. under ideal conditions.
  b. with limiting resources.
  c. at 6% per year.
  d. without deaths.

d 8. Overconsumption in developed nations affects
  a. energy consumption.
  b. global resources.
  c. waste production.
  d. All of these answers are correct.

Name _____ Class _____ Date _____

CHAPTER 13 REVIEW, CONTINUED

Understanding Trends Read the following statements, and choose the correct answer.

a 1. A J-curve represents a population that is
  a. increasing.
  b. decreasing.
  c. remaining the same.

c 2. As shown in Figure 13-7, the human population _____ around 1100 A.D.
  a. increased
  b. decreased
  c. remained the same

b 3. Populations that exceed the carrying capacity of their environment tend to
  a. increase.
  b. decrease.
  c. remained the same.

b 4. According to Figure 13-8, the amount of time needed by the human population to double _____ until about 1987.
  a. increased
  b. decreased
  c. remained the same

c 5. According to the theory of demographic transition, pictured in Figure 13-10, populations in Stage 1 tend to
  a. increase.
  b. decrease.
  c. remain the same.

b 6. During Stage 2 of a population's demographic transition, the death rate
  a. increases.
  b. decreases.
  c. remains the same.

c 7. When you are healthy, the growth rate of bacteria in your intestine tends to
  a. increase.
  b. decrease.
  c. remain the same.

c 8. As shown in Figure 13-5, populations that have reached the carrying capacity of their environment usually do not
  a. increase.
  b. decrease.
  c. remain the same.

a 9. Figure 13-12 shows that the population in developing countries is
  a. increasing.
  b. decreasing.
  c. remaining the same.

c 10. The population pyramid in Figure 13-13 shows that the odds of survival in developed countries tend to _____ over the first four decades of life.
  a. increase
  b. decrease
  c. remain the same

b 11. A population with a large number of emigrants and a low biotic potential is probably
  a. increasing.
  b. decreasing.
  c. remaining the same.

c 12. Biotic potential will _____ as a population increases in size.
  a. increase
  b. decrease
  c. remain the same

CHAPTER 13 REVIEW, CONTINUED

Short Answer Write the answers to the following questions in the spaces provided.

1. Explain the change in lifestyle that we often call the *agricultural revolution*.

   About 10,000 years ago, people began to switch from an itinerant, hunting and gathering way of life, to a more rooted and domesticated lifestyle centered around agriculture. Rather than move from place to place in search of food, people began to grow their own crops and raise domestic animals. People also learned to store food, which helped them support larger families. Agriculture increased the carrying capacity of the land, which led to an explosion of the human population over the past 1000 years.

2. Compare and contrast Stage 1 and Stage 3 of the demographic transition.

   During Stage 1 of the demographic transition, both birth and death rates are high, and the population grows slowly if at all. Therefore, the population is stable. In Stage 3, population growth is also stable, but it is because the birth rate has now fallen to equal the now low death rate. So the two stages are similar in that birth and death rates roughly cancel each other out in both stages. Stages 1 and 3 differ in that birth and death rates are both high in Stage 1, while they are both low in Stage 3.

3. Explain what an S-curve is and what it illustrates.

   An S-curve is the S-shaped portion of a population growth curve. It illustrates the growth of a population as it approaches the carrying capacity of its environment. It can be thought of as a J-curve that becomes flattened out at the top by the force of environmental resistance.

CRITICAL THINKING WORKSHEET

AGREE OR DISAGREE

CHAPTER 13

Agree or disagree with the following statements, and support your answer.

1. We could apply the term *environmental refugees* to other animal species as well as to humans.

   Accept any reasonable answer. Sample answer. Agree; animals must often flee an environmental disaster, whether natural or human-caused. Examples include birds flying away from the hot clouds of ash spewed from erupting volcanoes, water snakes slithering in search of new ponds and rivers during a drought, and bears wandering into towns after being displaced from their natural habitat.

2. Population-related problems are primarily the concern of developing countries.

   While many of the day-to-day population problems are felt most keenly by developing countries, the problems ultimately affect all countries, and most countries contribute to these problems as well. Developed countries do not usually have surging populations, but they do consume a much larger share of the world's resources than do developing countries. Developing countries tend to face the more immediately obvious problems associated with population growth, including air and water pollution, overcrowding, high infant mortality, and problems with food distribution.

3. There should be a limit on the amount of natural resources used per individual in developed countries.

   Accept any thoughtful response. Sample answer: Disagree; although this sounds like a good idea, it would be very difficult to enforce and would inevitably lead to bitterness and international disagreements. However, it might work for certain resources, such as metals and old-growth timber, whose removal is handled by a few large companies.

Page 98

Name _____ Class _____ Date _____

CRITICAL THINKING WORKSHEET

PROBLEM SOLVING

CHAPTER 13

Read the following problems, and solve them in the spaces provided.

Math hints: Calculators can help us work with exponents. Use decimals to represent portions of 100 percent. In other words, to represent a population with a 3 percent growth rate use 1.03. The 1 represents 100 percent of the current population, and the .03 represents the 3 percent growth rate. To calculate the percent that this population would grow over several years, take this growth rate to the power of the number of years in question, then multiply by 100. So to calculate the percent growth over five years, you would write $(1.03)^5 = 1.16 \times 100\% = 116\%$. If the population originally had 100 individuals, it would have 116 individuals after five years. Most calculators use a "y^x" key to calculate exponents. In this case, you would type 1.03, y^x, 5, = to get 1.16.

1. Six golden toads, three males and three females, begin a new population in a pond in the Peruvian Andes. The population doubling time in this new environment is 4 months. Assuming the population grows at its biotic potential, how long will it take the population to reach 192 organisms?

   $6 \times 2 \times 2 \times 2 \times 2 \times 2 = 192$; Five doublings times 4 months per doubling equals 20 months

2. On January 1, 1980, a country has 200 million people and an annual population growth rate of 8 percent. Over time, the growth rate falls, averaging 6 percent over the next 10 years. The growth rate then remains at 4 percent for the next 10 years. How large is the country's population on January 1, 2000? Show your work, and round your answer to the nearest million people.

   The simplest way to solve this problem is to use the average population growth rate during the 20-year period in question. The 8 percent growth rate does not figure in at all because it applied to an earlier period. The population averaged 6 percent for the first decade in question and 4 percent for the next, for an average of 5 percent over 20 years. Therefore, the total percent growth is $(1.05)20 = 2.653$, meaning that the population grew about 265 percent during the two decades. The final population size is 200 million $\times$ 2.653 = 531 million people.

Page 99

Name _____ Class _____ Date _____

CRITICAL THINKING WORKSHEET

READING COMPREHENSION AND ANALYSIS I

CHAPTER 13

Read the following passage, and answer the questions that follow.

Like all living things, people have an inherent tendency to multiply geometrically—that is, the more people there are the more people they tend to produce. In contrast, the supply of food rises more slowly, for unlike people, it does not increase in proportion to the existing rate of food production. This is, of course, the familiar relationship described by Malthus that led him to conclude that the population will eventually outgrow the food supply (and other needed resources), leading to famine and mass death. The problem is whether other countervailing forces will intervene to limit population growth and to increase food supply.

Barry Commoner*

1. Describe the relationship that Malthus noticed between the growth of the human population and its food supply.

   Malthus noticed that human population has a tendency to increase geometrically, while the supply of food rises more slowly. He predicted that this would eventually lead to the human population outgrowing its food supply.

2. What events have helped us avoid the mass death that Malthus predicted?

   Accept any reasonable response. Sample answer: Several countervailing forces have helped us elude a human population crash. One of the most important was the green revolution, which has allowed developing nations to produce enough grain to feed their populations. Another is the fact that nations have cooperated to distribute food so as to help other nations during times of drought and famine.

3. Is it reasonable to assume that these forces will enable us to avoid future population growth problems?

   Accept any thoughtful response. Sample answer: It all depends on the time frame involved. Over the short term, these forces will probably help stave off disaster. Over the long term, increasing the land's carrying capacity will not be enough by itself. Countless examples in nature have taught us that no population can increase indefinitely.

*From *Making Peace with the Planet* by Barry Commoner. Copyright © 1992 by Barry Commoner. Reprinted by permission of *The New Press*.

Name _____ Class _____ Date _____

READING COMPREHENSION AND ANALYSIS II

CHAPTER 13

Read the following passage, and answer the questions that follow.

People are perceived as poor if they eat millet (grown by women) rather than commercially produced and distributed processed foods sold by global agribusiness. They are seen as poor if they live in self-built housing made from natural material like bamboo and mud rather than in cement houses. They are seen as poor if they wear handmade garments of natural fiber rather than synthetics. Subsistence, as culturally perceived poverty, does not necessarily imply a low physical quality of life.

Vandana Shiva*

1. In your own words, what is the author saying in this passage?
The author is saying that people living off of the natural resources available in their environment may still be enjoying a high quality of life. She believes that one's quality of life is not determined by the extent to which one is separated from one's natural environment.

2. How might living at a subsistence level be the most adaptive response to the pressures people feel in many developing nations?
Living at a subsistence level ensures that one is able to at least meet the basic requirements for living, assuming they are still available in one's environment. Such a lifestyle also helps ensure that the nation's resources are not depleted as quickly as they might be otherwise.

3. Why do you think that many people believe that living at a subsistence level is equivalent to having a low quality of life?
Accept any thoughtful response. Sample answer: Many people believe this because, historically, it has been a "luxury" to live at anything but a subsistence level. Now that many people are able to afford more than the necessities of life, they think that living humbly is humble living.

*From "Development, Ecology, and Women" by Vandana Shiva from *Healing the Wounds*, edited by Judith Plant. Copyright © 1992 by Judith Plant. Reprinted by permission of *New Society Publishers*.

Name _____ Class _____ Date _____

TOWARD A SUSTAINABLE FUTURE

CHAPTER 14

Matching Match each example in the left column with the appropriate term from the right column.

c 1. public expresses views on policy — a. nonbinding
a 2. an unenforceable agreement — b. EIS
e 3. the world's first national park — c. hearings
f 4. blueprint for protecting the environment and promoting sustainable development — d. Earth Summit
d 5. conference to develop international agreements on the environment — e. Yellowstone
b 6. required by the federal government — f. Agenda 21

Concept Mapping

7. Complete the unfinished diagram below to illustrate the connections between the different components.

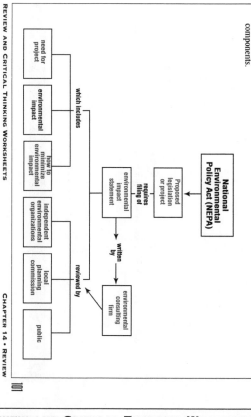

National Environmental Policy Act (NEPA) — requires filing of → Proposed legislation or project — written by → environmental consulting firm — reviewed by → independent environmental organizations, local planning commission, public

which includes: need for project, environmental impact — how to minimize environmental impact

Multiple Choice In the space provided, write the letter of the word or statement that best answers the question or completes the sentence.

a 1. Which tends to be most responsive to citizen input?
a. local government
b. state government
c. federal government
d. an international agency

b 2. Which is NOT an international environmental agreement?
a. Agenda 21
b. IWC
c. RAMSAR
d. Law of the Sea

d 3. We can use scientific knowledge and methods to
a. help make informed environmental decisions.
b. investigate nature.
c. influence policy.
d. All of these answers are correct.

c 4. Which group influences the type of development allowed in a particular area?
a. state agencies
b. planning and zoning boards
c. local governments
d. Both b and c are correct.

a 5. Which agreement seeks to eliminate wetland destruction?
a. RAMSAR
b. CITES
c. MARPOL
d. Agenda 21

d 6. Which problems require international cooperation?
a. overpopulation
b. loss of biodiversity
c. global warming
d. all of the above

a 7. How many days must the public have to comment on an Environmental Impact Report?
a. 45 days
b. 60 days
c. 90 days
d. none of the above

d 8. A sustainable future requires that we as individuals and as a society
a. learn about and continue to explore environmental science.
b. carefully consider a range of values.
c. take an active role in our communities.
d. All of these answers are correct.

Short Answer Write the answers to the following questions in the spaces provided.

1. Define and describe the purpose of an Environmental Impact Statement.
An environmental impact statement is a written assessment of the effects that a proposed law or project would have on the quality of the environment. An EIS states the need for a project, its impact on the environment, and how the impact could be minimized. EISs are published so the public may comment on the project. The agency in question must take public comments into consideration, which may result in modification of the project.

2. What can an individual do to affect environmental policy at the national level?
A person can contact members of Congress to express their opinions, or they can become a member of an organization dedicated to the preservation of the environment. Many organizations are in a position to lobby the government and influence legislation.

3. Explain two problems with the 1949 International Whaling Commission agreement.
One problem is that nations can opt out of a decision up to 90 days after the fact, such as France chose to do. A second problem is that although countries agree to quotas that cap the number of whales they can kill, the quotas do not specify species. A country could fill their quota with a species of whale that already had a low population, such as the blue whale.

4. How can environmental science help us make good environmental decisions?
Environmental science offers us a body of knowledge about the environment as well as methods to help us further understand how the environment functions. The knowledge and methods of environmental science enable us to make decisions based on an understanding of how the environment works and how it is likely to respond to our actions.

Name _____ Class _____ Date _____

CRITICAL THINKING WORKSHEET

ANALOGIES

CHAPTER 14

Mark the letter of the pair of terms that best completes the analogy shown. An analogy is a relationship between two pairs of words or phrases written as a:b::c:d. The symbol : is read *is to*, and the symbol :: is read *as*.

Example
keyboard : type ::
    a. plane : land
    b. dog : eat
 ✓ c. scissors : cut
    d. rock : hard
    e. grass : green

1. Voting : elected representative ::
    a. voting : local elections
 ✓ b. joining lobbying organizations : national policy
    c. reading EISs : planning
    d. organizing neighborhood meetings : state government
    e. driving : commuting

2. international agreements : nations ::
    a. lobbyists : national policy
    b. EIS : public
 ✓ c. game rules : baseball players
    d. public opinion : citizens
    e. legislation : policy

3. environmental science : environmental policy ::
    a. engineering : NASA
    b. thoughts : ideas
    c. education : high school
 ✓ d. medicine : health policy

4. values : decisions ::
    a. ideas : actions
    b. negotiations : agreements
    c. knowledge : ideas
    d. actions : future
 ✓ e. all of the above

5. international cooperation : environmental problems ::
    a. conferences : agreements
 ✓ b. teamwork : obstacles
    c. science : policy
    d. environment : sustainable
    e. local : national

6. Earth Day : environmental awareness ::
    a. garage sales : fliers
    b. ban : whaling
    c. EIS : environmental impact
 ✓ d. government : citizens
    e. ban : conservation

7. national policy : state policy ::
    a. state policy : national policy
    b. local policy : local politics
 ✓ c. state policy : local policy
    d. big fish : small fry
    e. all of the above

Name _____ Class _____ Date _____

CRITICAL THINKING WORKSHEET

AGREE OR DISAGREE

CHAPTER 14

Agree or disagree with the following statements, and support your answer.

1. Just as the state of California has created more rigorous air quality standards than are nationally mandated, an individual is obligated to do more to preserve the environment than is required by law.

    Accept any thoughtful answer. Sample answer: Agree; environmental regulations require the minimum amount of effort required not to make environmental problems worse. To move towards a sustainable future, people must work to improve the environmental situation, not just keep it from getting worse.

2. Wealthy nations should offer poor nations economic incentives to protect the environment.

    Accept any thoughtful answer. Sample answer: Agree; because wealthy nations are better able to meet the basic human requirements of food and shelter, it is their duty to help make the environment a priority for less fortunate nations as well as a priority for themselves. We will all benefit from a healthy environment; providing economic incentives to poorer nations is a worthwhile investment in the future.

3. Because it is so difficult for local communities to coordinate their efforts, action at the local level rarely has an effect on the environment.

    Disagree; the formation of the Hudson River greenway, in New York, was due to local government response to citizen input. Local governments make decisions about habitat preservation, population density, recycling, sewage treatment, and water quality. Even if their efforts are not coordinated (though they often are), local communities often have profound effects, both good and bad, on their local environments.

Name _____ Class _____ Date _____

REFINING CONCEPTS

CHAPTER 14

The statements below challenge you to refine your understanding of concepts covered in the chapter. Think carefully, and answer the questions that follow.

1. The United States and other highly developed nations have been vocal about the logging and burning of the rain forest in the Brazilian Amazon. Beyond the environmental implications—depletion of natural resources with no plan for renewal and an increase in global production of carbon dioxide—what makes this a controversial issue?

   Nations are very protective of their right to manage their own affairs. Also, during its development, the United States was free to do what it wished with its own natural resources and was not regulated by other countries. Trying to control the development of Brazil could be viewed as a hypocritical action.

2. How could the widespread study of environmental science affect government policy?

   Accept any thoughtful response. Sample answer: In a democracy, government policy is modeled after the values of the average citizen. The widespread study and understanding of environmental science affect values of the people. When people realize that a sustainable future is jeopardized, they might reevaluate their priorities and pressure elected policy makers to reflect their constituents' concerns in the nation's laws.

3. How do you think the EIS system might be made more effective?

   Accept any thoughtful response. Sample answer: EISs are prepared by environmental firms hired by the project planners. Project planners have an interest in keeping problems to a minimum. The environmental firm preparing the report is aware of this, and it can influence the outcome of the report. If the Environmental Impact Statements are all prepared by a federally regulated agency, this problem might be avoided.

Name _____ Class _____ Date _____

READING COMPREHENSION AND ANALYSIS I

CHAPTER 14

Read the following passage, and answer the questions that follow.

Earth Day 1970 represented a momentous breakthrough in the wall of cultural myopia that the Western World had put between itself and the Natural World. But it is also as though the rest of humanity arrived at the cosmological home of the Indigenous Peoples and, once there, failed to greet the host family properly.

José Barreiro*

1. What do you think the author meant by the idea of *cultural myopia?*

   Accept any thoughtful answer. Sample answer: Myopia means nearsightedness. The author means that the West has taken a near-sighted, or short, view of its relationship with the natural world. It has looked only at what it could extract in the short term rather than at how it could exist in harmony with nature over the long term.

2. What ideas might represent the cosmological home of the Indigenous Peoples?

   The cosmological home of the Indigenous Peoples includes a respect for the Natural World and the concept of a cooperative relationship with it.

3. Do you agree with the author's statement that the Western World "failed to greet the host family properly"? Why?

   Accept any thoughtful answer. Sample answer: Yes, the Western World seemed to embrace the idea that the Natural World should be respected without acknowledging that Indigenous Peoples had been acting as good stewards for thousands of years.

*From "Indigenous People Are the 'Miners' Canary' of the Human Family" by Jose Barreiro from *Learning to Listen to the Land,* edited by Bill Willers. Copyright © 1991 by Island Press. Reprinted by permission of *Alexander Hoyt Associates.*

Name _____ Class _____ Date _____

READING COMPREHENSION AND ANALYSIS II

CHAPTER 14

Read the following passage, and answer the questions that follow.

The debate over environmental challenges cannot be reduced to assigning blame. Patterns of consumption and resource use in the industrialized countries of the North are certainly responsible for much environmental degradation in both the North and South. But rapidly growing populations, whatever their levels of consumption, also place a greater burden on resources and the environment. Ensuring sustainability will require people to make changes, in both the way they think about their environment and how they live in it.

Nafis Sadik*

1. Do you think specific groups are blamed too often for environmental problems? Why?

   Accept any thoughtful answer. Sample answer: Yes, specific groups, such as industries or the industrialized consumers of the North, often take the blame for what might better be described as a human problem. Human societies allow them to abuse the environment and condone their environmental abuse by purchasing their goods and services.

2. Do you agree with the author's idea that changes in thinking and lifestyle are necessary to ensure sustainability? Why?

   Accept any thoughtful answer. Sample answer: Yes, without significant changes in the way humans think about the environment, the exploding populations of countries in Asia and Africa will inherit the destructive attitudes of northern industrialized nations. Once attitudes are changed, actions need to be modified to ensure sustainability.

3. How might environmental problems be handled differently if all people were required to take an environmental science course, such as this one?

   Accept any thoughtful answer. Sample answer: If every person was required to learn about humanity's relationship to the environment, massive attitude changes would be more likely to occur. Local actions could enable positive global changes.

*From "A Population Policy for the World" by Nafis Sadik in *Environment*, 2nd edition. Copyright © 1998 by **Saunders College Publishing, a division of Harcourt Brace & Company.** Reprinted by permission of the publisher.

108